OXFORD MASTER SERIES IN STATISTICAL, COMPUTATIONAL, AND THEORETICAL PHYSICS

OXFORD MASTER SERIES IN PHYSICS

The Oxford Master Series is designed for final year undergraduate and beginning graduate students in physics and related disciplines. It has been driven by a perceived gap in the literature today. While basic undergraduate physics texts often show little or no connection with the huge explosion of research over the last two decades, more advanced and specialized texts tend to be rather daunting for students. In this series, all topics and their consequences are treated at a simple level, while pointers to recent developments are provided at various stages. The emphasis in on clear physical principles like symmetry, quantum mechanics, and electromagnetism which underlie the whole of physics. At the same time, the subjects are related to real measurements and to the experimental techniques and devices currently used by physicists in academe and industry. Books in this series are written as course books, and include ample tutorial material, examples, illustrations, revision points, and problem sets. They can likewise be used as preparation for students starting a doctorate in physics and related fields, or for recent graduates starting research in one of these fields in industry.

CONDENSED MATTER PHYSICS

1. M.T. Dove: *Structure and dynamics: an atomic view of materials*
2. J. Singleton: *Band theory and electronic properties of solids*
3. A.M. Fox: *Optical properties of solids, second edition*
4. S.J. Blundell: *Magnetism in condensed matter*
5. J.F. Annett: *Superconductivity, superfluids, and condensates*
6. R.A.L. Jones: *Soft condensed matter*
17. S. Tautz: *Surfaces of condensed matter*
18. H. Bruus: *Theoretical microfluidics*
19. C.L. Dennis, J.F. Gregg: *The art of spintronics: an introduction*

ATOMIC, OPTICAL, AND LASER PHYSICS

7. C.J. Foot: *Atomic physics*
8. G.A. Brooker: *Modern classical optics*
9. S.M. Hooker, C.E. Webb: *Laser physics*
15. A.M. Fox: *Quantum optics: an introduction*
16. S.M. Barnett: *Quantum information*

PARTICLE PHYSICS, ASTROPHYSICS, AND COSMOLOGY

10. D.H. Perkins: *Particle astrophysics, second edition*
11. Ta-Pei Cheng: *Relativity, gravitation and cosmology, second edition*

STATISTICAL, COMPUTATIONAL, AND THEORETICAL PHYSICS

12. M. Maggiore: *A modern introduction to quantum field theory*
13. W. Krauth: *Statistical mechanics: algorithms and computations*
14. J.P. Sethna: *Statistical mechanics: entropy, order parameters, and complexity*
20. S.N. Dorogovtsev: *Lectures on complex networks*

Lectures on Complex Networks

SERGEY N. DOROGOVTSEV

University of Aveiro

and

Ioffe Institute, St Petersburg

OXFORD
UNIVERSITY PRESS

OXFORD
UNIVERSITY PRESS

Great Clarendon Street, Oxford OX2 6DP

Oxford University Press is a department of the University of Oxford.
It furthers the University's objective of excellence in research, scholarship,
and education by publishing worldwide in

Oxford New York

Auckland Cape Town Dar es Salaam Hong Kong Karachi
Kuala Lumpur Madrid Melbourne Mexico City Nairobi
New Delhi Shanghai Taipei Toronto

With offices in

Argentina Austria Brazil Chile Czech Republic France Greece
Guatemala Hungary Italy Japan Poland Portugal Singapore
South Korea Switzerland Thailand Turkey Ukraine Vietnam

Oxford is a registered trade mark of Oxford University Press
in the UK and in certain other countries

Published in the United States
by Oxford University Press Inc., New York

First published 2010

British Library Cataloguing in Publication Data

Data available

Library of Congress Cataloging in Publication Data

Data available

Typeset by SPI Publisher Services, Pondicherry, India
Printed in Great Britain
on acid-free paper by
CPI Antony Rowe, Chippenham, Wilts

ISBN 978–0–19–954892–7 (Hbk.)
978–0–19–954893–4 (Pbk.)

1 3 5 7 9 10 8 6 4 2

Preface

This text is a very concise modern introduction to the science of networks, based on lectures which I gave at several universities to students and non-specialists. My aim is to introduce a reader without serious background in mathematics or physics to the world of networks.

The term 'complex networks' is young. It came to use in the late 1990s when researchers from very distinct sciences—computer scientists, biologists, sociologists, physicists, and mathematicians—started to intensively study diverse real-world networks and their models. This notion refers to networks with more complex architectures than, say, a uniformly random graph with given numbers of nodes and links. Usually, in these complex architectures, hubs—strongly connected nodes—play a pivotal role. In this sense, the great majority of real-world networks are complex.

The field of complex networks is currently a very hot and attractive research area. The reader may ask: why all the fuss around networks in fundamental sciences like physics? I prefer the question: why are networks so interesting? The answer is not only the tremendous importance of the Internet and cellular networks. The point is that the geometry and structural organization of these and many other networks are very different from those of other well-studied objects—lattices. Networks and their function cannot be understood based on theories developed for finite-dimensional lattices, and a new vision is needed.

On the other hand, random networks are objects of statistical mechanics. So the course is essentially based on the standard apparatus of classical statistical physics. There are already several excellent popular science books and serious reference volumes on complex networks, including books on particular types of networks. The introductory lectures for beginners fill the existing gap between these two kinds of literature. The intended audience is mostly undergraduate and postgraduate students in physics and other natural science disciplines. There is some risk that inevitable oversimplification will only create an illusion of understanding. I believe however that this illusion is not too dangerous and may even be stimulating. Moreover, I suggest that the strict selection of material and discussion of recent results and fresh ideas will make this thin book useful, even for many specialists in networks. The reader who needs more detailed information and rigorous derivations can afterwards refer to more difficult reference books and original papers.

I am deeply indebted to my friends and colleagues in Portugal for their encouragement and advice, first and foremost to Anna Rozhnova,

Alexander Goltsev, Alexander Povolotsky, Alexander Samukhin, António Luís Ferreira, Fernão Vístulo de Abreu, João Gamma Oliveira, José Fernando Mendes, Gareth Baxter, Massimo Ostilli, Rui Américo da Costa, Zhang Peng, and Sooyeon Yoon. I would like to warmly thank Adilson Motter, Agata Fronczak, Alessandro Vespignani, Andre Krzywicki, Ayşe Erzan, Bartłomiej Wacław, Béla Bollobás, Bo Söderberg, Bosiljka Tadic, Bruno Goncalves, Byungnam Kahng, David Mukamel, Des Johnston, Dietrich Stauffer, Dmitri Krioukov, Dmitri Volchenkov, Doochul Kim, Florent Krzakala, Geoff Rodgers, Geoffrey Canright, Ginestra Bianconi, Hildegard Meyer-Ortmanns, Hyunggyu Park, Jae Dong Noh, János Kertész, Jorge Pacheco, José Ramasco, Kwang-Il Goh, László Barabási, Marián Boguñá, Mark Newman, Maksim Kitsak, Martin Rosvall, Masayuki Hase, Matteo Marsili, Matti Peltomaki, Michael Fisher, Michel Bauer, Mikko Alava, Kim Sneppen, Kimmo Kaski, Oliver Riordan, Olivier Bénichou, Pavel Krapivsky, Peter Grassberger, Piotr Fronczak, Romualdo Pastor-Satorras, Santo Fortunato, Sergey Maslov, Shlomo Havlin, Sid Redner, Stefan Bornholdt, Stephane Coulomb, Tamás Vicsek, Zdzislaw Burda, and Zoltán Toroczkai for numerous helpful and stimulating discussions and communications. Finally, my warmest thanks to the superb editorial and production staff at Oxford University Press for their invaluable guidance, encouragement, and patience.

Aveiro
July 2009

S.N.D.

Contents

First steps towards networks

A network (or a graph) is a set of nodes connected by links.[1] In principle, any system with coupled elements can be represented as a network, so that our world is full of networks. Specific networks—regular and disordered lattices—were main objects of study in physics and other natural sciences up to the end of the 20th century. It is already clear, however, that most natural and artificial networks, from the Internet to biological and social nets, by no means resemble lattices. The path to understanding these networks began in St Petersburg in 1735 with a mathematical problem formulated on a very small graph.

[1] The terms 'vertices' and 'edges' are more standard in graph theory.

1.1 Euler's graph

This small undirected graph (Fig. 1.1) with multiple links was considered by legendary Swiss born mathematician Leonhard Euler (1707–1783). Young Euler was invited to St Petersburg in 1727 and worked there until his death, with a 25-year break (1741–1766) in Berlin. In 1735 Euler made what is now regarded as the birth point of graph theory: he solved the Königsberg bridge problem. The structure of all possible paths within Königsberg in Euler's time is represented in the form of a graph in Fig. 1.1. The nodes of the graph are separate land masses in old Königsberg, and its links are the bridges between these pieces of land. Could a pedestrian walk around Königsberg, crossing each bridge only once? In other words, is it possible to walk this graph passing through each link only once? Euler proved that such a walk is impossible.[2]

[2] Much later, in 1873, Carl Hierholzer proved that a walk of this kind exists if and only if every node in a graph has an even number of links.

In graph theory the total number of connections of a node is called *degree* (it is sometimes called connectivity in physics). Consequently, Euler's graph has three nodes of degree 3 and one of degree 5. According to another definition, a *simple graph* does not have multiple links and loops of length 1. Otherwise, the graph is a *multi-graph*. Thus Euler's graph is a multi-graph. Degree is a local characteristic. Any description of the structure of an entire network or of its parts is essentially based on two notions: a path and a loop. A *path* is an alternating sequence of adjacent nodes and links with no repeated nodes. A *cycle* (in graph theory) or a *loop* (in physics) is a closed path where only the start and end nodes coincide. Note that it is the presence of loops in Euler's graph that makes the Königsberg bridge problem fascinating and profound.

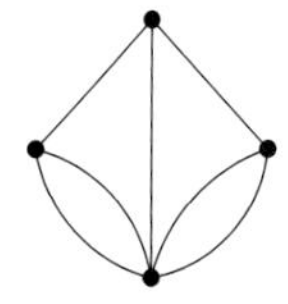

Fig. 1.1 Euler's graph for the Königsberg bridge problem. The undirected links of this graph are seven bridges of old Königsberg connecting four separate land masses—nodes: Kneiphof island and the banks of the Pregel river. As Euler proved in 1735, there is no walk on this graph that passes through each link only once.

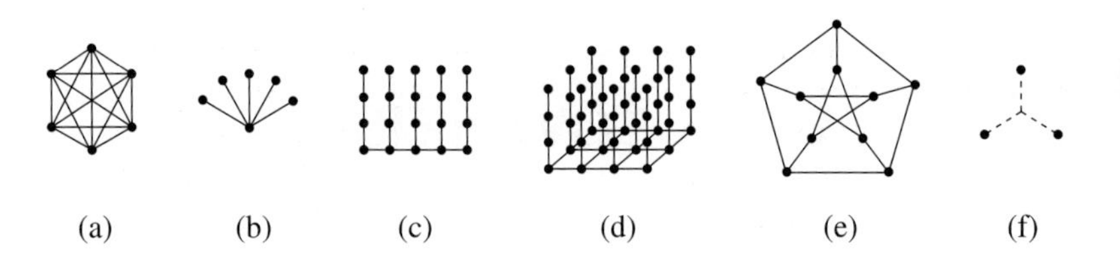

Fig. 1.2 (a) Fully connected or complete graph. (b) Star. (c) Comb graph. (d) Brush. (e) Petersen graph which is the $(3,5)$ cage graph, where 3 is the degree of its nodes and 5 is the length of the shortest cycle. (f) The simplest hypergraph: three nodes interconnected by a single hyperedge.

Graphs without loops—*trees*—are usually more simple to analyse. For example, a one-dimensional chain is a tree. The numbers of nodes N (which we call the 'network size') and links L of a tree satisfy a simple relation $L = N - 1$.

1.2 Examples of graphs

A few simple graphs are shown in Fig. 1.2. A complete graph, Fig. 1.2 (a), which is widely used in exactly solvable models in physics, has all nodes interconnected. In a star graph (which is the most compact tree), Fig. 1.2 (b), the maximum separation between nodes is 2. Combs and brushes, containing numerous chains, are shown in Figs. 1.2 (b) and (c), respectively. Due to the chains, random walks on these graphs essentially differ from those on lattices. The next example—the Petersen graph in Fig. 1.2 (e)—is one of the so-called *cage graphs*. These graphs are regular in the sense that each node in the graph has the same number of connections. A (q, g)-cage is a graph with the minimum possible number of nodes for a given node degree q and a given length g of the shortest cycle. In synchronized systems the cage graph architectures provide optimal synchronization. The notion of a graph can be generalized. In one of the direct generalizations—*hypergraphs*—generalized links ('hyperedges') connect triples, quadruples, etc. of nodes, see Fig. 1.2 (f).

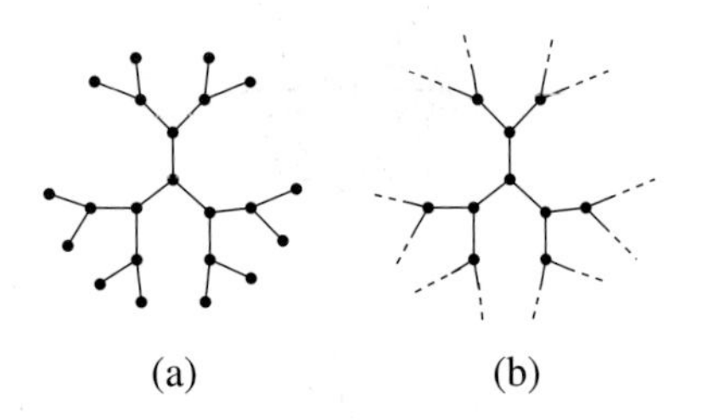

Fig. 1.3 (a) 3-regular Cayley tree. (b) Bethe lattice. Notice the absence of a border.

The next two important regular graphs, a *Cayley tree* and a *Bethe lattice*, shown in Fig. 1.3, will be extensively discussed in these lectures. These are very different networks. A Cayley tree has a boundary, which contains a finite fraction of all nodes—dead ends—and a centre (a root). A Bethe lattice is obtained from an infinite Cayley tree by formal exclusion of dead ends. As a result, all nodes in a Bethe lattice are equivalent, so there is neither a boundary nor a centre. To get rid of boundaries, physicists often treat these graphs as containing infinite loops.

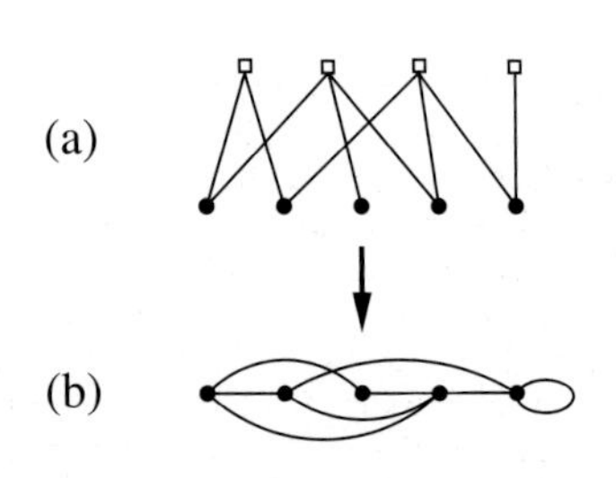

Fig. 1.4 A bipartite graph (a) and one of its one-mode projections (b).

Collaboration and many other networks may have not one but two types of nodes. For example a network of scientific coauthorships contains nodes–authors and nodes–papers. Each scientific paper in this graph is linked to all of its authors. As a result we have a *bipartite graph*, shown in Fig. 1.4. These networks are actually hypergraphs, where a 'node–paper' together with its connections plays the role of a 'hyperedge'. A one-mode projection of a bipartite graph, explained in Fig. 1.4, is less informative. Many empirical maps of networks are only one-mode projections of as yet unexplored real multi-partite networks.

1.3 Shortest path length

A distance ℓ_{ij} between two nodes i and j in a network is the length of the shortest path between them through the network. Two characteristics describe the separation between nodes in an entire network: the mean internode distance in the network and its diameter. The *mean internode distance* $\overline{\ell}$ (also mean geodesic distance) is the average of ℓ_{ij} over all those pairs of nodes (i, j) between which there is at least one connecting path. (Note that, in general, networks will contain disconnected parts.) The *diameter* ℓ_D is the maximum internode distance in a network. We will demonstrate that in many large networks, there is no great difference between these two quantities.

It is the dependence of $\overline{\ell}$ or ℓ_D on network size N that is particularly important for characterization of network architectures. In networks with a compact structure, which we discuss, $\overline{\ell}(N)$ grows with N slower than in more loose structures—lattices.

1.4 Lattices and fractals

In finite-dimensional regular and disordered lattices (these lattices are supposed to have no long-range bonds), the size dependence $\overline{\ell}(N)$ is power-law,

$$\overline{\ell} \sim N^{1/d}. \tag{1.1}$$

Here d is the dimensionality of a lattice—an integer number. In contrast, *fractals* (Fig. 1.5 explains this notion) may have non-integer dimensionalities. In fractals, $\overline{\ell} \sim N^{1/d_f}$, where d_f is called a fractal or Hausdorff dimension. Note that there is no great difference between finite-dimensional lattices and fractals in respect of the dependence $\overline{\ell}(N)$: they are both 'large worlds'. For example, for a two-dimensional lattice of 10^{12} nodes, $\overline{\ell} \sim 10^6$. Only when d or d_f tend to infinity does this dependence become non-power-law. Note that the fractal dimension can be found even if (large) N is fixed. Simply count down the number of nodes $n(\ell)$ within a distance ℓ from a given node. In fractals (and lattices) this number is $n \sim \ell^{d_f}$.

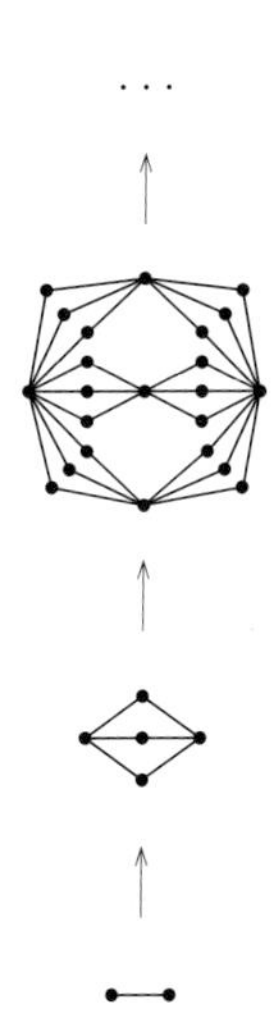

Fig. 1.5 This transformation generates a fractal of dimension $d_f = \ln 6/\ln 2 = 2.585\ldots$. At each step, every link is substituted by the cluster of six links. So, while the diameter of the graph (here it is the distance from its left node to the right one) is doubled, the total number L of links increases by a factor of six. If t is a step number, then $\overline{\ell} \sim \ell_D = 2^t$ and $L = 6^t$, and consequently $N \sim L = \overline{\ell}^{\,\ln 6/\ln 2}$.

For the sake of comparison, let us estimate $n(\ell)$ for the q-regular Bethe lattice and Cayley tree, Fig. 1.3. As is common, instead of the node degree q, we use another number—*branching* $b = q - 1$. Then $n = 1 + q(1 + b + b^2 + \ldots + b^{\ell-1}) \cong qb^{\ell-1}$, where we assume that n is large. Thus, $n \sim b^\ell$, in contrast to lattices and fractals, and so, for a Cayley tree, we have

$$\overline{\ell} \sim \frac{\ln N}{\ln b}, \tag{1.2}$$

which grows with N much slower than for any finite-dimensional lattice.

In this respect, one may say that Cayley trees and Bethe lattices are infinite-dimensional objects—'small worlds'. If, for example, a Cayley tree has 10^{12} nodes of degree, say, 5, then $\overline{\ell} \sim 10$, which is dramatically smaller than in the previous example for a two-dimensional lattice.

Generally, the term *small-world phenomenon* refers to a slower growth of $\overline{\ell}(N)$ than any positive power of N (and to a more rapidly growing $n(\ell)$ than any power of ℓ). The networks showing this phenomenon are called *small worlds*. Most of the explored real-world networks, which we will discuss, have compact architectures of this kind.

1.5 Milgram's experiment

The small-world phenomenon was first observed in a social network. In 1967 prominent social psychologist Stanley Milgram (1933–1984) performed a seminal experiment for measuring distances in a network of acquaintances in the United States.[3] The question was: how many intermediate social links separate two randomly selected (and geographically separated) individuals?

The idea of the experiment was elegant, see Fig. 1.6. Milgram chose two locations: Omaha, Nebraska and Boston.[4] A target person was chosen at random in Boston. A large enough number of randomly selected residents of Omaha received a letter with the following instructions:

(i) If you know the target person 'on a personal basis' (his/her name and address were enclosed), send the letter directly to him/her.

(ii) Otherwise mail a copy of this instruction to your 'personal' acquaintance (someone you know on a first name basis) who is more likely than you to know the target person.

An essential fraction of letters approached the target, after passing through only, on average, 5.5 social links; which is a surprisingly small number. This is what is known as the 'six degrees of separation'. One may think that the real shortest path length should be even smaller, since the experiment revealed only a small fraction of all possible chains between starting persons and the target.

It is dangerous however to believe sociologists too much: (i) they have to work with poorly defined and subjective material, (ii) they have to use poor statistics. The details of the experiment and the resulting number, the 'six degrees', were criticised,[5] but nobody denies the essence of Milgram's observation—the impressive smallness of the world of social relations.[6]

1.6 Directed networks

In directed networks, at least some fraction of connections are directed. It seems that the first extensively studied nets of this type were networks of citations in scientific papers.[7] The nodes of a citation network are scientific papers, and the directed links are citations of one paper within another (Fig. 1.7). New links in the citation networks emerge only between new nodes and already existing ones; new connections between existing nodes are impossible (one cannot update an already published paper). In graph theory networks of this kind are called *recursive graphs*.

[3] Milgram's paper in *Psychology Today* with the results of his experiment was entitled 'The small world problem' [125].

[4] In fact, Milgram made two attempts. The first one, with starting points in Wichita, Kansas and a target person in Sharon, Massachusetts, resulted in only three finished chains, but the second attempt turned out to be more successful.

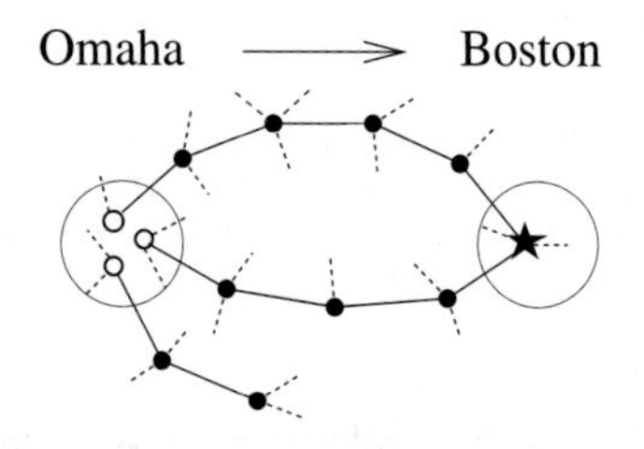

Fig. 1.6 How Stanley Milgram scanned a net of acquaintances in the United States. Notice that some chains of acquaintances were broken off.

[5] For an intelligent critique of the results of Milgram's experiment see Kleinfeld [113]. Note one of Kleinfeld's arguments: 'our desire to believe we live in a "small, small world"'.

[6] 36 years after Milgram his experiment was repeated on a greater scale by using the modern opportunity of email (Dodds, Muhamad, and Watts 2003) [70]. Volunteers started 24 163 chains aimed at reaching 18 target persons in 13 countries. Only 384 (!) of the chains were completed, which indicates that the global social world is rather disconnected. On the other hand, the successful chains turned out to be an average of about 4 links, i.e. even less than 'six degrees'.

[7] Price (1965) [152].

Furthermore, all links in a network of citations have the same direction—to older papers. This is valid, of course, for publications in paper form, that is in printed journals and in books. In contrast, papers in many electronic archives may be updated. I can update my old works in the http://arXiv.org electronic archive and change their lists of references to cite more recent papers.[8] So, some links in the citation networks of these electronic archives may be oppositely directed.

[8] The http://arXiv.org is one of the largest electronic archives, used mostly by physicists but also mathematicians and computer scientists. Most papers on the statistical mechanics of networks can be found in the cond-mat and physics sections of this archive.

1.7 What are random networks?

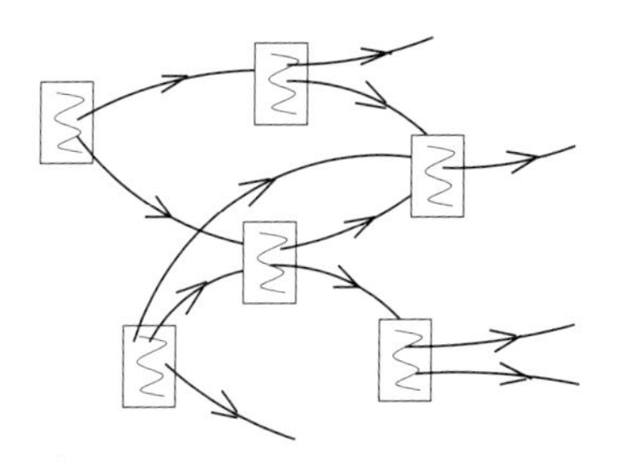

Fig. 1.7 Network of citations in scientific papers.

Even if we ignore the directedness of connections, the apparently random network in Fig. 1.7 differs from the graphs shown in Figs. 1.2 and 1.3. But then, what is a random network from the point of view of a physicist or a mathematician? Note that, strictly speaking, the notion of randomness is not applicable to a single finite graph. Indeed, by inspecting this finite graph, one cannot find whether it was generated by a deterministic algorithm or by a non-deterministic one. In the spirit of statistical physics, a random network is not a single graph but a *statistical ensemble*. This ensemble is defined as a set of its members—particular graphs—where each member has its own given probability of realization, that is its *statistical weight*.[9] By this definition, a given random network is some graph with one probability, another graph with another probability, and so on. To obtain some quantity, characterizing a random network, in principle we should collect the full statistics for all members of the statistical ensemble. To obtain the mean value of some quantity for a random network, we average this quantity over all members of the ensemble—over all realizations—taking into account their statistical weights.[10]

[9] More rigorously, the statistical weights are proportional to the realization probabilities. However, the proportionality coefficient is an arbitrary constant.

[10] Note the difference between the three kinds of scientist. As a rule, empirical researchers and experimenters collect statistics for a single realization of a random network. Scientists using numerical simulations (computer experiments) investigate a few or a relatively small number of realizations. Theorists consider all, or at least all essential, members of the statistical ensemble of a random network.

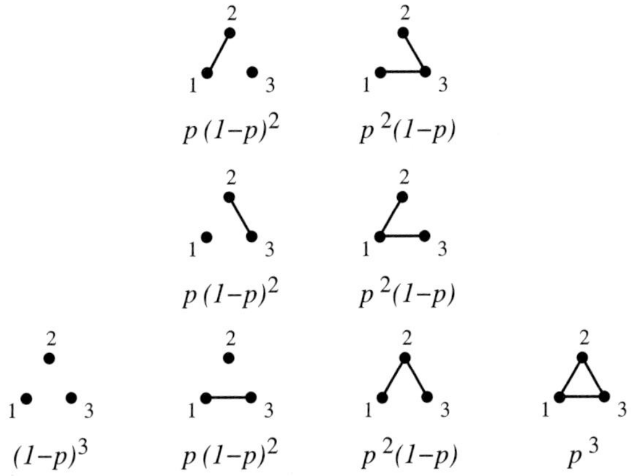

The first example of a random graph is a classical random graph model, shown in Fig. 1.8. This is the $G_{N,p}$ or Gilbert model defined as follows. Take a given number N of labelled nodes, say $i = 1, 2, 3, \ldots, N$, and interlink each pair of nodes with a given probability p. If $N = 3$, this gives eight possible configurations with the realization probabili-

$p(1-p)^2$ $p^2(1-p)$

$p(1-p)^2$ $p^2(1-p)$

$(1-p)^3$ $p(1-p)^2$ $p^2(1-p)$ p^3

Fig. 1.8 The Gilbert model of a random graph (the $G_{N,p}$ model) for N=3 with realization probabilities represented for all configurations. All graphs in each column are *isomorphic*, that is they can be transformed into each other by relabelling their nodes.

ties shown in the figure. Note that this graph is 'labelled' (has labelled nodes). As in classical statistical mechanics, where particles are distinguishable (i.e., can be labelled), networks are usually considered to be labelled, which is important for the resulting ensemble.

Physicists divide statistical ensembles into two classes—equilibrium and non-equilibrium—which correspond to equilibrium and non-equilibrium systems. This division is also relevant for random networks. For example, the ensemble presented in Fig. 1.8 is equilibrium—its statistical weights do not evolve. In non-equilibrium (evolving) ensembles, statistical weights of configurations vary with time, and the set of configurations may also vary. Growing networks are obviously non-equilibrium. However, even among networks with a fixed number number of nodes, one can find non-equilibrium nets.

Suppose now that the number of nodes in a random network approaches infinity. Then, as a rule, the statistics collected for one member of the ensemble almost surely coincides with the statistics for the entire ensemble—*self-averaging* takes place. In other words, a relative number of ensemble members with non-typical properties is negligibly small. It turns out that the self-averaging property is very common in disordered systems. So the features of many large, but finite individual graphs can be accurately described in terms of statistical ensembles. It is technically easier for a theoretical physicist to analyse a statistical ensemble than a single graph, and so the self-averaging is really useful.

1.8 Degree distribution

The degree distribution $P(q)$ is the probability that a randomly chosen node in a random network has degree q:

$$P(q) = \frac{\langle N(q) \rangle}{N} . \tag{1.3}$$

Here $\langle N(q) \rangle$ is the average number of nodes of degree q in the network, where the averaging is over the entire statistical ensemble. We assume that the total number of nodes in each member of the ensemble is the same, $N = \sum_q \langle N(q) \rangle$. An empirical researcher, who studies a single graph, say graph g, measures the frequency of occurrence of nodes with degree q in this graph: $P_g(q) = N_g(q)/N$. Here $N_g(q)$ is the number of nodes of degree q in graph g. This quantity is also usually called a degree distribution. $P_g(q)$ approaches $P(q)$ in the infinite network limit.

The degree distribution is the simplest statistical characteristic of a random network, and it is usually only the first step towards the description of a net. Remarkably, in many situations knowledge of the degree distribution is sufficient for the understanding of a network and the processes taking place on it. In principle, the entire degree distribution is significant: its low- and high-degree parts are important for different network properties and functions. In classical random graphs such as shown in Fig. 1.8, degree distributions decay quite rapidly, $P(q) \sim 1/q!$

for large q (see the next lecture). All their moments $\sum_q q^n P(q)$ are finite even as the network size approaches infinity, and so the mean degree $\langle q \rangle = \sum_q qP(q)$ is a typical scale for degrees. There are practically no strongly connected hubs in these networks.

In contrast, numerous real-world networks, from the Internet to cellular nets, have slowly decaying degree distributions, where hubs occur with noticeable probability and play essential roles. Higher moments of the degree distributions of these networks diverge if we tend the size of the network to infinity. A dependence with power-law asymptotics $P(q) \sim q^{-\gamma}$ at large q gives a standard example of a slowly decaying degree distribution.[11] The power-law distributions are also called scale-free and networks with these distributions are called *scale-free networks*. This term implies the absence of a typical node degree in the network.[12]

[11] The value of the moment $\sum_q q^n P(q)$ here is determined by the upper limit of the sum. In an infinite network, this limit approaches infinity. So, if exponent $\gamma \leq n+1$, then the nth and higher moments of the distribution diverge.

[12] More strictly, the term 'scale-free' refers to the following property of a power-law distribution $q^{-\gamma}$. A rescaling of q by a constant, $c \to cq$, only has the effect of multiplication by a constant factor: $(cq)^{-\gamma} = c^{-\gamma}q^{-\gamma}$.

1.9 Clustering

Clustering is about how the nearest neighbours of a node in a network are interconnected, so it is a non-local characteristic of a node. In this respect clustering goes one step further than degree. The clustering coefficient of a node is the probability that two nearest neighbours of a node are themselves nearest neighbours. In other words, if node j has q_j nearest neighbours with t_j connections between them, the local clustering coefficient is

$$C_j(q_j) = \frac{t_j}{q_j(q_j-1)/2}\,, \tag{1.4}$$

see Fig. 1.9. When all the nearest neighbours of node j are interconnected, $C_j = 1$; when there are no connections between them, as in trees, $C_j = 0$. The number t_j is the total number of triangles—loops of length 3—attached to the node, and so the clustering refers to the statistics of small loops—triangles—in a network. Importantly, most real-world networks have strong clustering.

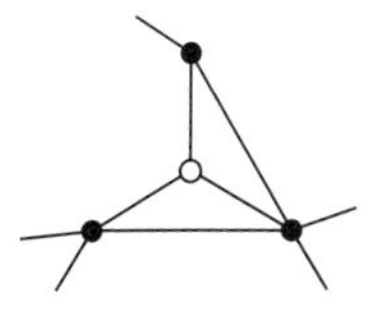

Fig. 1.9 The clustering coefficient of the central node equals 2/3.

In general, the clustering coefficient of a node depends on its degree. Empirical researchers often present their data on degree-dependent clustering by using an averaged quantity—the mean clustering coefficient of a node of degree q—that is $\overline{C}(q) = \langle C_j(q) \rangle$. Two different less informative integral characteristics of network clustering are traditionally used. The first is the *mean clustering* of a network, which is the average of the local clustering coefficient, eqn (1.4), over all nodes, $\overline{C} = \langle t_j/[q_j(q_j-1)/2] \rangle = \sum_q P(q)\overline{C}(q)$. The second characteristic—the *clustering coefficient* C of a network or *transitivity*—allows one to find the total number of loops of length 3 in the network.[13] The clustering coefficient of a network is defined as

$$C = 3\,\frac{\text{the number of loops of length 3 in a network}}{\text{the number of connected triples of nodes}}\,. \tag{1.5}$$

A triple here is a node and two of its nearest neighbours.[14] A 3-loop

[13] The notion of clustering was adapted from sociology, where it is usually called transitivity.

[14] One can easily find that the number of connected triples of nodes equals $\sum_i q_i(q_i-1)/2 = N(\langle q^2 \rangle - \langle q \rangle)/2$.

consists of three triples, which explains the coefficient 3. The denominator gives three times the maximum possible number of loops of length 3. One can easily see that C is also the ratio of the average numerator of expression (1.4) and its average denominator, $C \equiv \langle t_j \rangle / \langle q_j(q_j-1)/2 \rangle$. Compare this with the definition of mean clustering. If $\overline{C}(q)$ is independent of degree q, then the mean clustering and the clustering coefficient coincide, $\overline{C} = C$.

1.10 Adjacency matrix

Networks are naturally represented in matrix form. A graph of N nodes is described by an $N \times N$ adjacency matrix $\hat{a}$ whose non-zero elements indicate connections between nodes.[15] For undirected networks, a non-diagonal element a_{ij} of an adjacency matrix is equal to the number of links between nodes i and j, and so the matrix is symmetric. A diagonal element a_{ii} is twice the number of loops of length 1 attached to node i. The factor 2 here is clear: each 1-loop plays the role of a double connection for a node. As a result, the degree of node i is $q_i = \sum_j a_{ij}$.

[15] In a random network, each of the members of the statistical ensemble is represented by its own adjacency matrix.

Any structural characteristic of a network can be expressed in terms of the adjacency matrix. See, for example, the expression for the total number T of triangles in a graph without 1-loops:

$$T = \frac{1}{6}\sum_i (\hat{a}^3)_{ii} = \frac{1}{6}\mathrm{Tr}\,\hat{a}^3 . \qquad (1.6)$$

Here Tr denotes the trace of a matrix—the sum of its diagonal elements.[16] This formula leads to a compact expression for the clustering coefficient.

[16] Check that the total number of links, $L = K/2 \equiv \sum_i k_i$, in this graph is

$$L = \frac{1}{2}\sum_{ij} a_{ij} = \frac{1}{2}\sum_i (\hat{a}^2)_{ii} = \frac{\mathrm{Tr}\,\hat{a}^2}{2}.$$

Numerical calculations with adjacency matrices of large networks require huge memory resources. Fortunately, one can often avoid using adjacency matrices. The point is that real-world networks and their models are typically *sparse*. That is, the numbers of connections in these networks are much smaller than in complete graphs: $L \ll N^2$, i.e. $\langle q \rangle \ll N$. In 1999, in the WWW, for example, the average number of outgoing and incoming hyperlinks per web page was about eight. Therefore the great majority of matrix elements in the adjacency matrices of these networks are zeros. So, instead of an adjacency matrix $N \times N$, it is better to use a set of N vectors, $i = 1, 2, \ldots, N$, where the components of vector i are the labels of the nearest neighbours of node i. This takes up much less memory, $\langle q \rangle N \ll N^2$.

Classical random graphs

2

In this lecture we give an insight into the simplest and most studied random networks—classical random graphs.

2.1 Two classical models

In 1951–1952, applied mathematicians Ray J. Solomonoff and Anatol Rapoport published a series of papers in the *Bulletin of Mathematical Biophysics*, which did not attract much attention at the time [165, 166]. It is in these papers that the $G_{N,p}$ or Gilbert model, as it is now known, of a random graph (see Fig. 1.8) was introduced.[1] Later this basic model was rediscovered by mathematician E. N. Gilbert (1959). The notation $G_{N,p}$ indicates that this is a statistical ensemble of networks, G, with two fixed parameters: a given number of nodes N in each ensemble member and a given probability p that two nodes have an interconnecting link.

[1] The terms Bernoulli or binomial random graph are also relevant. The term 'binomial' is explained by the binomial form of statistical weights in this model, see Fig. 1.8.

In the second half of the 1950s outstanding Paul Erdős and Alfréd Rényi introduced another random graph model and actually established random graph theory as a field of mathematics [87, 88]. The *Erdős–Rényi random graph* is a statistical ensemble whose members are all possible labelled graphs of given numbers of nodes N and links L, and all these members have equal statistical weight. This random graph is also called the $G_{N,L}$ model—a statistical ensemble G of graphs with two fixed parameters for each member of the ensemble: (i) a given number of nodes N and (ii) a given number of links L.[2] This is a special case of a general construction extensively exploited in the science of networks. The idea of this basic construction is to build an ensemble whose members are all possible graphs satisfying given restrictions, and all these members are realized with equal probability—uniformly randomly. One can say, this is the maximally random network that is possible under the given restrictions. One can also say that a network of this kind satisfies a given constraint but is otherwise random. Figure 2.1 shows an example—a small Erdős–Rényi network of 3 nodes and 1 link. Compare this small ensemble with that of the $G_{N=3,p}$ model in Fig. 1.8 and note a clear difference between these two ensembles. Remarkably, this difference turns out to be negligibly small in sufficiently large sparse networks.

[2] Mathematicians usually use the notations $G_{n,p}$ and $G_{n,M}$ for the ensembles $G_{N,p}$ and $G_{N,L}$, respectively.

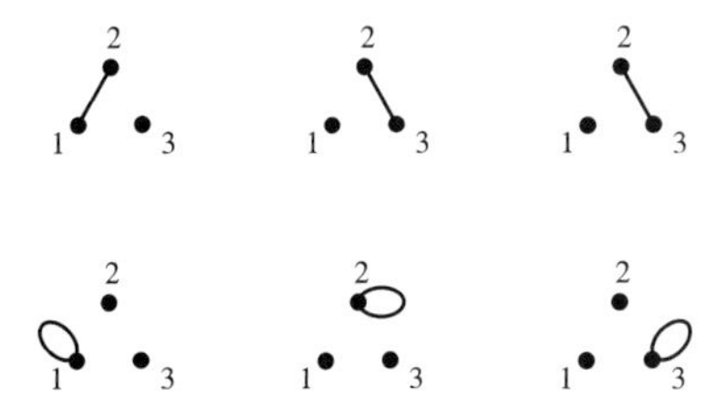

Fig. 2.1 The Erdős–Rényi random graph of 3 nodes and 1 link, which is $G_{N=3,L=1}$. All six graphs have equal statistical weight. Draw all configurations of the $G_{N=3,L=2}$ random graph.

Let us discuss this point in more detail. In simple terms, statistical mechanics is based on two kinds of ensembles—*canonical and grand canonical ensembles*. In all members of the first ensemble, the number of particles is equal and fixed. In the second, grand canonical ensemble, the chemical potential is fixed, and the numbers of particles are different in different ensemble members. In statistical physics, these two ensembles—in our case two ways to define a random system—usually become equivalent as the number of particles approaches infinity. In random graphs links actually play the role of particles. So the $G_{N,L}$ and $G_{N,p}$ models correspond to the canonical and grand canonical ensembles, respectively. Suppose that $N \to \infty$, and the networks are sparse. Then one can show that these two random networks approach each other if $p = L/[N(N-1)/2]$.[3] In that sense, these two models are so close that they are called together 'classical random graphs' or even simply 'random graphs'.[4] Moreover, the term 'Erdős–Rényi model' sometimes refers to both of these ensembles. The reader may be surprised that in contrast to the Gilbert model, the Erdős–Rényi graph contains multiple connections and 1-loops, see Fig. 2.1. Then how can these models be equivalent? The explanation is that multiple connections and 1-loops in a large sparse Gilbert graph are not important—we will see that there are few of them.

[3] More precisely, the statistical characteristics of typical members of these two ensembles converge. In particular, a large majority of members in the Gilbert ensemble have only relatively small deviations in the numbers of links from $L = pN(N-1)/2$. Note that $N(N-1)/2$ is the total number of links in the complete graph of N nodes.

[4] One can even say that classical random graphs are maximally random networks with a given mean degree $\langle q \rangle$ of a node.

Physicists know that analysis of the grand canonical ensembles is technically easier than for the canonical ones. In this respect the Gilbert model has an advantage. Let us obtain the degree distribution of this network using intuitive arguments. A node in this random graph can be connected to each of the other $N-1$ nodes with probability p. Then combinatorics immediately results in the binomial form of the probability that q of these $N-1$ links are present:

$$P(q) = C^q_{N-1} p^q (1-p)^{N-1-q}\,, \tag{2.1}$$

which is the degree distribution of the finite graph. Here $C^q_n = n!/[q!(n-q)!]$ is the binomial coefficient. One can obtain this exact formula strictly, averaging over the ensemble. The resulting mean degree of a node is $\langle q \rangle = p(N-1)$. When the network is large ($N \to \infty$) while $\langle q \rangle$ is finite (i.e. $p \to \text{const}/N$), the binomial distribution (2.1) approaches the Poisson one:

$$P(q) = e^{-\langle q \rangle} \langle q \rangle^q \frac{1}{q!}\,. \tag{2.2}$$

In this limit the Gilbert and Erdős–Rényi models are equivalent, and so this Poisson distribution is valid for all classical random graphs. The extremely rapid decay of the distribution is determined by the factorial denominator. We have mentioned that the degree distributions of practically all interesting real-world networks decay much slower.

Importantly, the degree distributions in these random graphs completely describe their architectures. Indeed, links in these networks are distributed independently. The only restriction is the fixed mean degree of a node. Therefore a node in a classical random graph 'does not know' the statistics of connections of any other node. In that sense, even connected nodes are statistically independent. Random graphs of this kind are called *uncorrelated networks*, which is one of the basic notions in this field.[5] We will consider general uncorrelated networks in detail in the following lectures.

[5] In particular, the absence of correlation means factorization of the joint distributions of random variables. For example, let $P(q, q')$ be the probability that one node has degree q and another one has degree q'. Then in an uncorrelated network, $P(q, q') = f(q)f(q')$, where the function $f(q)$ is completely determined by the degree distribution, see Lecture 5.

2.2 Loops in classical random graphs

As was mentioned, the large sparse classical random graphs have few loops. What does this mean? Let us first discuss small loops. It is easy to find the clustering coefficient and so the total number of loops of length 3 in a large classical random graph. Let the graph be of N vertices with a mean degree $\langle q \rangle$. Recall that the clustering coefficient of a node is the probability that two nearest neighbours of the node are themselves nearest neighbours. In our case this probability is $\langle q \rangle/(N-1) \cong \langle q \rangle/N$. So the clustering is

$$C = \overline{C} = \frac{\langle q \rangle}{N} \,. \tag{2.3}$$

Note the equality $C = \overline{C}$ and the independence of the clustering coefficient of a node from its degree. This is, of course, the case for any uncorrelated network. So the clustering coefficient for an infinite sparse classical random graph approaches zero. That is, clustering in these networks is only a finite size-effect. Let us compare the clustering coefficients of real-world networks with those of the classical random graphs with the same numbers of nodes and links. For example, these days (2009) the map of routers in the Internet contains about half a million nodes—routers—or even more. The Internet grows exponentially, and so this number changes rapidly. The mean number of connections of a node in this network is about 10. The clustering coefficient is about 0.1. For a classical random graph of the same size, with the same mean degree, formula (2.3) gives $C \sim 10^{-5}$, which is four orders of magnitude lower than in the Internet! In the 1990s, when exploration of real-world networks was in its early stages, many empirical studies highlighted this tremendous difference, thus showing how far classical random graph models stray from reality.

Formula (2.3) allows us to find the total number of triangles in a classical random graph:[6]

$$\mathcal{N}_3 = \langle q \rangle^3/6 \,. \tag{2.4}$$

That is, the number of triangles in a sparse classical random graph does not depend on its size. This number is finite even if these graphs are infinite. Similarly, one can find the number of loops of an arbitrary length L. This number also does not depend on L: $\mathcal{N}_L \cong \langle q \rangle^L/(2L)$ if L is smaller than the diameter of the network, which we expect to be of the order of $\ln N$. Thus any finite neighbourhood of a node in these random graphs almost certainly does not contain loops. In that sense, these networks are *locally tree-like*, which is a standard term. We will use this convenient feature extensively. On the other hand, there may be plenty of long loops of length exceeding $\ln N$; namely, $\ln \mathcal{N}_L \sim N$ if $L \gg \ln N$. Obviously, these loops cannot spoil the local tree-like character.

One may observe another important object, *cliques*. A clique is a fully connected subgraph. For example, a loop of length 3 provides us with a 3-clique. Since there are so few loops in these networks, one can see that

[6] This expression is obtained in the following way. $\mathcal{N}_3 = \frac{1}{3}C(\#\text{ of triples})$, see Section 1.9. The total number of connected triples of nodes is $N(\langle q^2 \rangle - \langle q \rangle)/2$. Finally, we use the fact that for the Poisson distribution the second and first moments are related:

$$\langle q^2 \rangle - \langle q \rangle = \langle q \rangle^2 .$$

3-cliques are the maximum possible. The bigger cliques in large sparse classical random graphs are almost entirely absent.

There is a similar random graph construction, even more simple than the Erdős–Rényi model. This is the *random regular graph.* All N vertices of this graph have equal degrees. The construction is similar to the Erdős–Rényi model. The random regular graph is a statistical ensemble of all possible graphs with N vertices of degree q, where all the members are realized with equal probability. In other words, this is a maximally random regular graph.[7] This graph also has loops of various lengths, including loops of length 1, as in the Erdős–Rényi graph. Their number is $\mathcal{N}_L \cong (q-1)^L/(2L)$, and so these networks also have a locally tree-like structure. In that sense, an infinite random regular graph approaches the Bethe lattice with the same node degree.

[7] Note that a random regular graph does not belong to the category of classical random graphs.

2.3 Diameter of classical random graphs

Now we can immediately exploit the local tree-like character of simple random networks. We essentially repeat our derivation of the diameter of a Bethe lattice from Section 1.4. The only difference being that, in a random tree one should substitute a branching b in the derivation for the mean branching $\overline{b}$ of a link. So the number z_n of the n-th nearest neighbours of a node grows as $\overline{b}^{n-1}$. This exponential growth guarantees that the number of nodes S_n which are not farther than distance n from a given node is of the same order, $\overline{b}^n$. Of course, this tree ansatz fails when S_n is already close to the size of a network, that is when n is close to the diameter. It turns out, however, that if we ignore the presence of loops even in this range of n and estimate $\overline{b}^{\overline{\ell}} \sim N$, we get a non-essential error in the limit of large N. So the resulting mean internode distance is $\overline{\ell} \cong \ln N/\ln \overline{b}$ at large N. This result is valid for all uncorrelated networks.

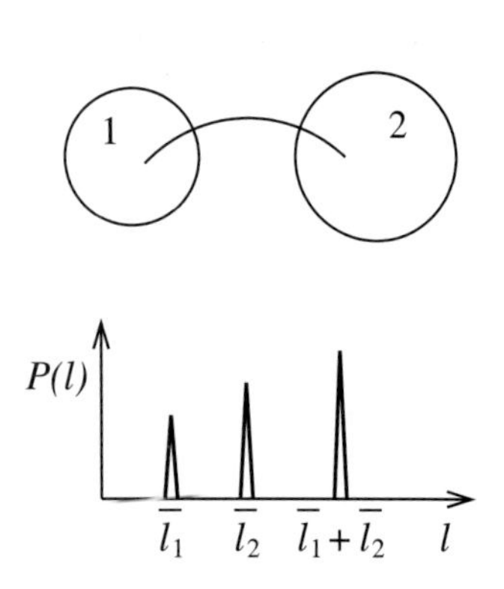

Fig. 2.2 A network consisting of two large modules—small worlds—interconnected by a single link has a wide distribution $\mathcal{P}(\ell)$ of internode distances. Indeed, let the mean separations between nodes inside modules 1 and 2 be $\overline{\ell}_1$ and $\overline{\ell}_2$, respectively. Then two nodes in different modules are separated by approximately $\overline{\ell}_1+\overline{\ell}_2$ links. So the distribution has three peaks: at $\overline{\ell}_1$, $\overline{\ell}_2$, and $\overline{\ell}_1+\overline{\ell}_2$. If, however, the number of connections between the modules is any finite (nonzero) fraction of the total number of links, then all three peaks merge into a single narrow peak.

The exponential growth, $z_n \sim \overline{b}^n$, has another remarkable consequence. Since $z_{\overline{\ell}} \sim N$, a finite fraction of nodes in these networks are at distance $\overline{\ell}$ from each other. That is, the width of distribution of internode distances $\mathcal{P}(\ell)$ in these networks is finite, even if these nets are infinitely large. In other words, nodes in infinite networks are almost surely mutually equidistant. The width $\delta\ell$ of the distribution $\mathcal{P}(\ell)$ is finite only in specific network models. However, the relatively narrow internode distance distributions, $\delta\ell \ll \overline{\ell}$, are typical for a wide range of small worlds.[8] (For a rare counterexample, a specific small world without this property, see Fig. 2.2.) This is why the mean internode distance in large networks is close to the diameter (the maximum separation between nodes). Note that mutual equidistance is realized only in very large networks, where the mean separation of nodes is much greater than 1. Even the largest real-world nets (in the WWW, for example, $\overline{\ell} \sim 10$, which is not that great) still have rather wide internode distance distributions.

[8] For comparison, one can easily estimate that for d-dimensional lattices, $\delta\ell/\overline{\ell} \sim 1-2^{-d}$, and so this distribution is wide if d is finite.

For a random q-regular graph, branching equals $q-1$, and so readily

$\bar{\ell} \cong \ln N/\ln(q-1)$. To obtain the diameter of a classical random graph, we must first find the average branching of its links. For this purpose, we will use a remarkably general relation for networks. Consider an arbitrary undirected graph of N nodes—a single realization, not an ensemble. This graph may include bare nodes and be simple or multiple. Let the numbers of nodes of degree $q = 0, 1, 2, \ldots$, be $N(q)$. So $N = \sum_q N(q)$, and the frequency of occurrence of nodes of degree q is the ratio $N(q)/N$. The total degree of the graph (double the number of links) equals $\sum_q qN(q) = N\langle q\rangle$. Clearly, $N(q)$ nodes of degree q attract $qN(q)$ stubs ('halves of links') of total number $N\langle q\rangle$. Then the frequency with which end nodes of a randomly chosen link have degree q, is $qN(q)/(N\langle q\rangle)$. In other words, select a link at random, choose at random one of its end nodes, then the probability that it has q connections equals $qN(q)/(N\langle q\rangle)$.

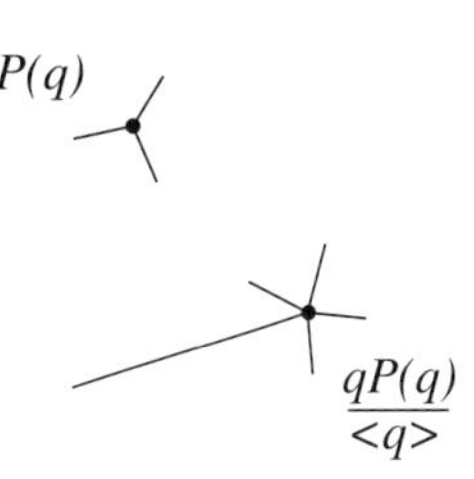

Fig. 2.3 End nodes of a randomly chosen link in a network have different statistics of connections from the degree distribution of this network.

For random networks, this important statement is formulated as follows. *In a random network with a degree distribution $P(q)$, the degree distribution of an end node of a randomly chosen link is $qP(q)/\langle q\rangle$*, see Fig. 2.3. Therefore the connections of end nodes of links are organized in a different way from those of randomly chosen nodes. We will use this fact extensively in the following sections. In particular we have $\langle q^2\rangle/\langle q\rangle$ for the average degree of an end of a randomly chosen link, which is greater than the mean degree of a node $\langle q\rangle$ in the network. The mean branching is the average value of the number of connections of an end of a randomly chosen link, minus one—the link itself. So it is $\bar{b} = (\langle q^2\rangle/\langle q\rangle) - 1$. Recall that for the Poisson distribution, $\langle q^2\rangle - \langle q\rangle = \langle q\rangle^2$. Therefore in classical random graphs, an average branching is equal to a mean degree, $\bar{b} = \langle q\rangle$. Finally we arrive at the famous formula

$$\bar{\ell} \cong \frac{\ln N}{\ln\langle q\rangle} \tag{2.5}$$

for the mean separation between nodes in a classical random graph and for its diameter. Interestingly, this formula provides a reasonably good estimate for mean internode distance in numerous real-world networks, even if they differ strongly from classical random graphs. For example, for the map of routers, $N \sim 2 \times 10^5$ and $\langle q\rangle \sim 3$, formula (2.5) gives $\bar{\ell} \approx 11$, which is close to an empirical value.

2.4 The birth of a giant component

In our discussion of diameters we missed one thing. In general, a graph may consist of different, separate parts—clusters. All nodes inside each of these parts—*connected components*—have connecting paths running within them.[9] Deriving formula (2.5) for a diameter we actually supposed that a random graph consists of a single connected component. This is certainly not true when a mean degree $\langle q\rangle$ approaches zero and the random graph consists of bare nodes. Already in 1951 Ray Solomonoff and Anatol Rapoport had discovered that when $\langle q\rangle$ exceeds 1,

[9] We assume here that a network is undirected.

infinite classical random graphs contain a large connected component including a finite fraction of nodes. This is a so-called *giant connected component.* Physicists, working in the field of condensed matter, are familiar with a close analogy—a *percolation cluster* [169]. Remove at random a fraction of nodes from an infinite conducting lattice, so that a fraction p of nodes are retained. Then below some critical value of p the lattice is split into a set of finite disconnected clusters, and the system is isolating. On the other hand, a current can flow—'percolate'—from one border of the lattice to another if p is above this critical concentration—a percolation threshold p_c. The current flows on an infinite percolation cluster of connected nodes, which is quite similar to a giant connected component. The situation in classical random graphs is as follows, see Fig. 2.4. When a classical random graph has many links, $\langle q \rangle > 1$, it consists of a single giant connected component and, also, numerous 'finite connected components'. On the other hand, if $\langle q \rangle < 1$, the giant connected component is absent and there are only plenty of finite connected components.

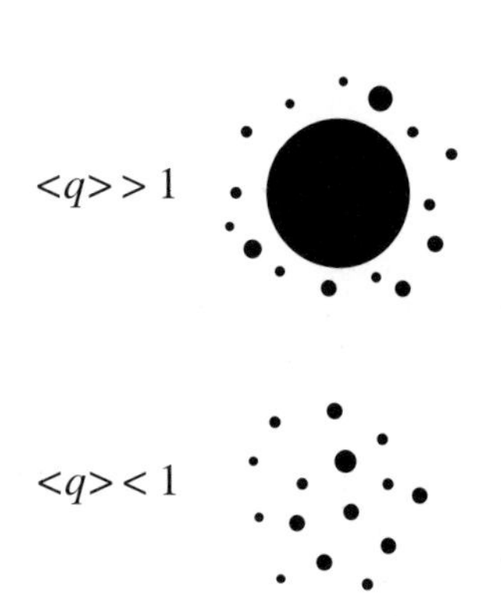

Fig. 2.4 The organization of a classical random graph. The filled circles show connected components. When mean degree $\langle q \rangle$ is above 1, the graph contains a giant connected component, which is absent if $\langle q \rangle < 1$.

This qualitative picture is generic for networks. The general properties of a network are primarily determined by whether or not a giant connected component is present. So the first question about any network should be about the presence of this component. Strictly speaking, a giant connected component is well defined for infinite networks. What about finite nets? We must inspect the dependence on network size, N. If the number of nodes in a connected component grows proportionally to N, then it is treated as 'giant'. In contrast, finite connected components practically do not grow with N.

Solomonoff and Rapoport found that a giant connected component in classical random graphs emerges exactly when a mean degree $\langle q \rangle$ surpasses 1. This happens without a jump, see Fig. 2.5. In that sense the birth of a giant connected component is a continuous phase transition, where $\langle q \rangle = 1$ is the critical point.[10] This is the main structural transition in a network, where network characteristics dramatically change. Note that all these changes take place in the regime of a sparse network, in which the number of connections is low compared to the maximal possible number. Furthermore, Fig. 2.5 demonstrates that a giant connected component approaches the size of a classical random graph still being in a sparse regime. In particular, the relative size of a giant connected component is already above 99% at $\langle q \rangle = 5$. Thus main qualitative changes in the architecture of networks are in the narrow region $\langle q \rangle \ll N$. Remarkably, most studied real-world networks are sparse.

[10] In physics, if there is no jump of an order parameter (in our case, the relative size of a giant component) at the critical point, then it is a continuous phase transition.

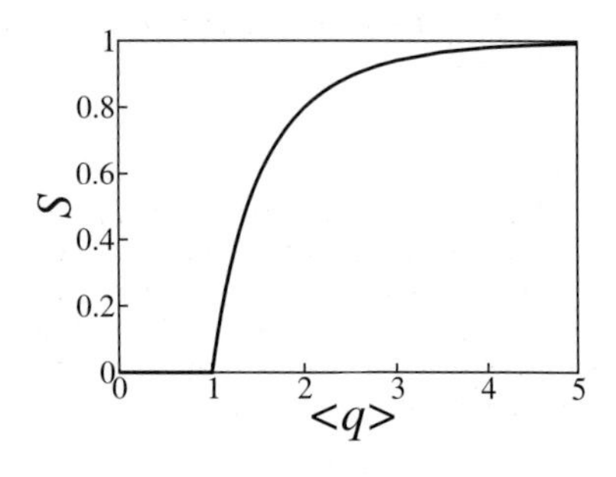

Fig. 2.5 The relative size of a giant connected component in a classical random graph versus the mean degree of its nodes. Near the birth point, $S \cong 2(\langle q \rangle - 1)$.

Returning to formula (2.5) for the diameter, we may conclude that it certainly fails close to the birth point of a giant connected component. We will discuss this special point in detail in the following lectures. On the other hand, when a giant component contains a reasonably large fraction of nodes in a classical random graph, deviations from relation (2.5) are negligible.

2.5 Finite components

After Erdős and Rényi, mathematicians spent a few decades studying the statistics and structure of connected components in the classical random graphs. In simple terms, the overall picture looks to be as follows. Let the size N of a classical random graph tend to infinity. We already know about a giant component. What about 'finite' ones? Note these inverted commas. The point is that this standard term may be confusing. It turns out that some 'finite connected components' still grow with N but extremely slowly, much slower than a giant connected component (whose size is proportional to N). Let us first stay away from the critical point $\langle q \rangle = 1$ in either of the two phases: in the 'normal phase', without a giant component, or in the phase with a giant component. Then the biggest 'finite' connected component, the second biggest, the third, the i-th biggest, with i being any finite number, all of these components have sizes of the order of $\ln N$. The total number of components grows with N, and so most of the components are really finite as $N \to \infty$. In any case, the sizes of all of these components are much smaller than that of the giant one.

Let us now move to the the critical point, where a giant component is still absent. At this point, the biggest connected component, the second, the third, the i-th biggest, with i being any finite number, all of these components are of the order of $N^{2/3}$.[11] It is important that this size, $N^{2/3}$, is much smaller than N but much bigger than $\ln N$.

[11] Rigorously speaking, this is valid within a so-called scaling window, where the deviation from the critical point, $|\langle q \rangle - 1|$, is less than, say, $N^{-1/3}$, Bollobás (1984) [39].

The statistics of connected components are remarkably different at the critical point and away from it. In the normal phase and in the phase with a giant component, these distributions have a rapid exponential decay. In contrast, at the critical point, the size distribution of components $\mathcal{P}(s)$ decays slowly, as a power law:

$$\mathcal{P}(s) \sim s^{-5/2} . \tag{2.6}$$

The sum $\sum_s s\mathcal{P}(s) \sim \sum_s s s^{-5/2}$ converges. Thus the mean size $\langle s \rangle$ of a finite connected component is finite at any value of $\langle q \rangle$, including the birth point of a giant connected component. Figure 2.6 (see the lower curve) shows that the finite components are mostly very small—one or two nodes.

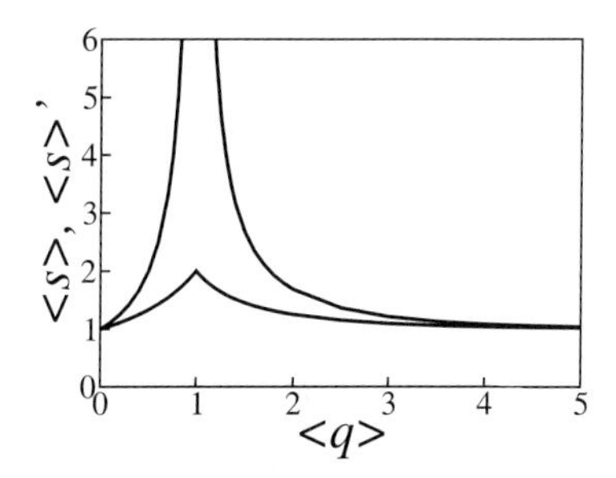

Fig. 2.6 The lower curve: the mean size (number of nodes) $\langle s \rangle$ of a finite component in a classical random graph versus the mean degree of its nodes. The upper curve: the mean size $\langle s \rangle'$ of a finite component to which a randomly chosen node belongs as a function of mean degree. $\langle s \rangle'$ diverges at the critical point as $1/|\langle q \rangle - 1|$, while $\langle s \rangle$ is finite.

We see from Fig. 2.6 that the dependence of $\langle s \rangle$ on $\langle q \rangle$ has only a not particularly impressive cusp at the critical point. On the other hand, another, related average demonstrates a real critical singularity. Choose at random a node in a network. What is the probability that this node is in a connected component of s nodes? Clearly, this probability $\mathcal{P}'(s)$ is proportional to the product $s\mathcal{P}(s)$. At the birth point of a giant connected component, $\mathcal{P}'(s) \sim s^{-3/2}$, so it has an infinite first moment. Thus the average size $\langle s \rangle'$ of a finite connected component to which a (randomly chosen) node belongs diverges at the critical point. The upper curve in Fig. 2.6 shows the full dependence of $\langle s \rangle'$ on $\langle q \rangle$. In the critical region, $\langle s \rangle' \cong 1/|\langle q \rangle - 1|$, which strongly resembles the famous Curie–Weiss law for susceptibility in physics.[12]

[12] The Curie–Weiss law works for interacting systems with a second-order phase transition provided that the mean-field theory is valid. For example, near a critical temperature in ferromagnets the magnetization

$$M(T, H = 0) \propto \sqrt{T_c - T}$$

in a zero applied field. A susceptibility $\chi(T, H)$ characterizes the response of the magnetization to a small addition of an applied field: $H \to H + \delta H$. That is, $\chi(T, H) \equiv \partial M(T, H)/\partial H$. According to the Curie–Weiss law, the susceptibility at a zero applied field is

$$\chi(T, H = 0) \propto 1/|T - T_c|.$$

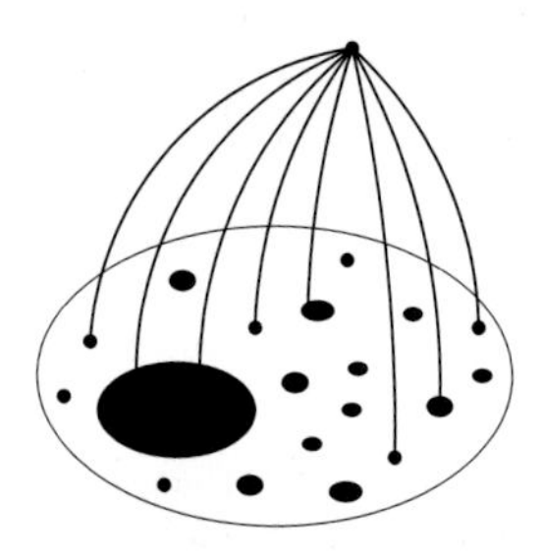

Fig. 2.7 An additional node attached to randomly chosen nodes in a network increases the giant connected component and so plays the role of an external field. The black spots are connected components in a network.

One may even say that the average $\langle s \rangle'$ plays the role of 'susceptibility' in graph theory and in percolation problems. Figure 2.7 explains this analogy. Let us add a node to an arbitrary network and link it to a number n of randomly chosen nodes. Suppose first that this number is a finite fraction of the total number of nodes N. Then, thanks to the connections of the 'external' node, we obtain a giant connected component even if this component is absent in the original network. Clearly, this specific attachment plays the role of an applied field which increases the size $M = SN$ of a giant connected component—similarly to the magnetic field changing magnetization in a ferromagnet. We intentionally use the same notation for the size of a giant connected component as for magnetization in physics. Similarly to magnetization (the ferromagnetic order parameter) the size of a giant connected component plays the role of an order parameter in networks. The 'susceptibility' may now be introduced as an increase in the size of a giant component in response to a small addition to the number of 'external' links $n \to n + \delta n$. The 'zero field susceptibility' is then $\partial M(n)/\partial n|_{n=0}$. One can easily see that this is exactly the average size $\langle s \rangle'$, since the attachments are to randomly selected nodes.

To conclude this lecture, we emphasize that the qualitative picture described here for classical random graphs is of a surprisingly general nature going even beyond graph theory and the science of networks. In this respect, this is a zero model of a random network, but it is not a toy model.

Small and large worlds

3

3.1 The world of Paul Erdős

The fantastic productivity of Paul Erdős and his travelling life (Erdős had a reputation as a mathematician-pilgrim) resulted, in particular, in an incredible number of coauthors. Over 500 mathematicians had the privilege of being coauthors with Erdős. So Erdős plays the role of a hub in the network of collaborations between mathematicians. Nodes in this net are the mathematicians (authors), and undirected connections are coauthorships in at least one publication. We have already mentioned that this is actually a one-mode projection of a bipartite collaboration network. In total, the network of collaborations between mathematicians includes about 337 000 authors, connected by 496 000 links.

It is a great honour to be Erdős' coauthor, but also it is a honour to be a coauthor of Erdős' coauthor (in reality, this honour is shared by a lot of mathematicians), and so on. These grades of the 'closeness' to Erdős are classified by *Erdős numbers*. The Erdős number of Erdős is 0, that of his coauthors is 1, of coauthors of Erdős' coauthors is 2, *etc.* In short, the Erdős number is the shortest path length between a mathematician and Erdős. Figure 3.1 shows the numbers of mathematicians with various Erdős numbers. One can easily see that this is actually the distribution of internode distances in this network. Note that this world of Erdős has a crucially different structure from the Erdős–Rényi model. Indeed, a rapidly decreasing Poisson degree distribution does not admit hubs with 500 in graphs with such a small number of connections. By using the formula for the Poisson distribution, the reader can find the probability that in a classical random graph with the same numbers of nodes and links as in the mathematics collaboration graph, at least one node has 500 connections. This probability is of the order of 10^{-934}.

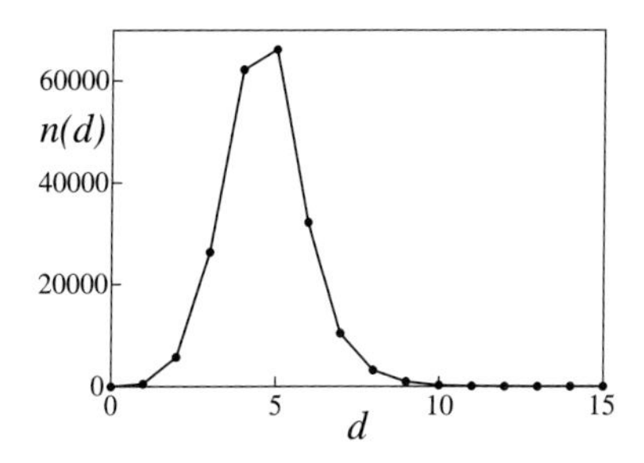

Fig. 3.1 The number of mathematicians with Erdős number d in 2001. This plot was made by using data from the web page of the Erdős Number Project, http://www.oakland.edu/grossman/erdoshp.html.

One can see from Fig. 3.1 that the average distance of a mathematician from Erdős is about 5—only five steps/coauthorships from greatness! For comparison, formula (2.5) gives the mean internode distance 12 for a classical random graph with the same number of nodes and links. Thus this network is even more compact than the classical random graph, though the difference is not dramatic. Note that the distribution in this figure is not narrow. The diameter equals 15, which is three times bigger than the mean distance between nodes. This is in contrast to the mutual equidistance of nodes in infinite small worlds, which we discussed. This is not that surprising, since the mean internode distance is not much greater than 1.

One should note that the described small world of Paul Erdős is a very typical collaboration network. All these networks are small worlds. Now, what about the largest directed net—the WWW?

3.2 Diameter of the Web

In 1999 Reka Albert, Hawoong Jeong, and Albert-Laszlo Barabási, physicists from the University of Notre Dame, measured what they called 'the diameter' of the WWW [5]. In fact, this was the average length of, what is important, the directed path between two pages of the WWW. Albert, Jeong, and Barabási succeeded in collecting data from a small part of the WWW, namely the nd.edu domain of the University of Notre Dame. Then, how could such limited data be used for obtaining the mean internode distance of the large Web?

The researchers used the following approach. Although the complete map of the nd.edu domain contained only 325 729 pages and 1 469 680 hyperlinks, these numbers were large enough to expect that the organization of connections in the domain is close to that in the entire Web. The architecture of the nd.edu was approximately reproduced in a set of even smaller model networks of different sizes. These small networks could easily be generated and studied numerically. The only features of the real WWW they reproduced were the distributions of incoming and outgoing connections of nodes, but this was sufficient.[1] It was easy to measure the mean shortest-directed-path length for each of these mini-Webs. The resulting size dependence $0.35 + 0.89 \ln N$ was extrapolated to the estimated size of the Web in 1999, $N \approx 800\,000\,000$, which gave the mean internode distance 19. In the spirit of our calculations of $\overline{\ell}$ in classical random graphs, we can easily estimate the average length of the shortest directed path in a directed network: $\overline{\ell}_d \approx \ln N / \ln\langle q_o \rangle$. Here $\langle q_o \rangle$ is a mean out-degree, that is the mean number of outgoing links of a node.[2] Substituting the above numbers to this estimate we obtain $\overline{\ell}_d \approx 14$ for the network of 800 000 000 nodes, which is not very far from the result of Albert, Jeong, and Barabási.

Later direct measurements made by a large group of computer scientists from AltaVista, IBM, and Compaq resulted in a close value, 16, for the part of the WWW (about 200 000 000 pages) observed by the search engine AltaVista [44]. In other words, a web navigator would have to make only about 20 'clicks' on average to reach a web page, following hyperlinks. Thus, in respect of internode distances, the large Web is a very small object—a small world.

[1] As a model of the WWW these authors used a maximally random network with a given distribution of incoming and outgoing links of nodes.

[2] By definition, the in-degree q_i of a node is the number of its incoming links, the out-degree q_o is the number of outgoing connections.

3.3 Small-world networks

The collaboration networks and the WWW are only two important examples among an infinity of real networks showing the small-world feature. We also mentioned that the great majority of real-world networks have strong clustering. The values 0.1, 0.2, 0.3, etc. for a clustering coefficient are very typical. This combination—the small-world phenomenon and the strong clustering—seemed to be completely incomprehensible in the late 1990s, when the only reference model widely used for comparison was a classical random graph. Unfortunately, classical random graphs have very weak clustering. Ten years later, it is clear that there exist plenty of easy ways to get a strongly clustered small world. The nature of strong clustering (and, in general, of numerous loops) is not considered to be a serious problem now. Nonetheless, it was the desire to understand the high values of clustering coefficient in the empirical data that inspired sociologist Duncan Watts and applied mathematician Steven Strogatz (1998) to propose an original model, interconnecting themes of networks and lattices [181]. This very popular model has significantly influenced the development of the field.

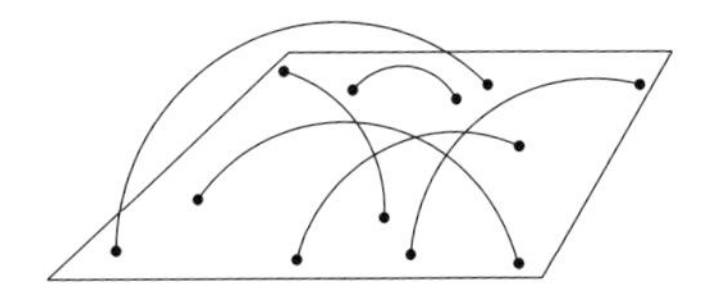

Fig. 3.2 The idea of a small-world network. Long-range shortcuts—links connecting randomly chosen nodes—are added to a d-dimensional lattice. Together with the nodes and connections within the mother lattice (which are not shown) these links form a small-world network.

The basic idea of Watts and Strogatz was as follows. Suppose somebody, who knows only lattices and the Erdős–Rényi model, wants to construct a network with the small-world feature and numerous triangles. The classical random graphs demonstrate the small-world phenomenon but have few triangles. On the other hand, many lattices have numerous triangles (if their unit cells include triangles of bonds) but the lattices have no small-world feature. Then let us combine a lattice with many triangles and a classical random graph with the small-world feature. The combination of the 'geographically short-range' connections and long-range shortcuts has clear roots in real life: find examples from communications, sociology, etc. Technically, Watts and Strogatz connected pairs of randomly chosen nodes in a lattice by links—'shortcuts', see Fig. 3.2. Thanks to these long-range shortcuts, even nodes, widely separated geographically within the mother lattice, have a chance to become nearest neighbours. Clearly, the shortcuts make the resulting networks more compact than the original lattice. Watts and Strogatz called networks of this kind *small-world networks*. This term is used even if a mother lattice is disordered, or it has no triangles. The clustering here is of secondary importance. Indeed, what is so special about loops of length 3? Loops of length 4, 5, etc. are no less important.

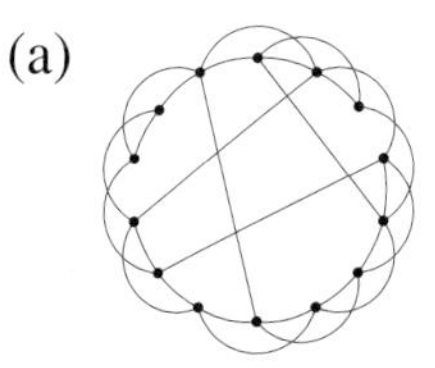

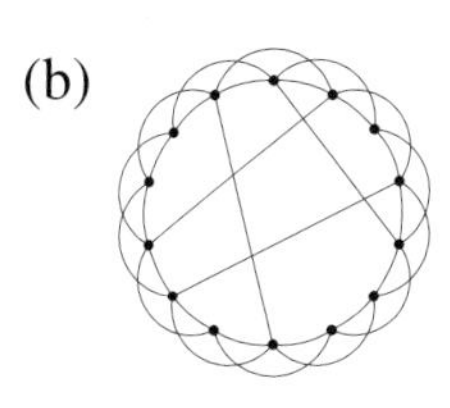

Fig. 3.3 The original Watts–Strogatz model (a) with rewiring of links [181], and its variation (b) with addition of shortcuts (b) [132]. Notice numerous 3-loops within the mother lattice.

Figure 3.3 (a) shows the original Watts–Strogatz network where randomly chosen links of a mother lattice were rewired to randomly selected nodes. The rewiring produces the same effect as the added long-range shortcuts, Fig. 3.3 (b). Figure 3.3 (b) also explains why we have superposition of a lattice and a classical random graph. Indeed, if we let the connections within the mother lattice be absent, then the shortcuts and the nodes form a classical random graph. The resulting small-world network is already 'a complex one'.

The problem is: what does happen when the number of shortcuts increases? It is easy to see that even a single long-range shortcut sharply diminishes the average internode distance $\overline{\ell}$. For example, the first shortcut added to a one-dimensional lattice (as in Fig. 3.3) may reduce $\overline{\ell}$ by half! So, the influence of shortcuts on the shortest path lengths in a small-world network is dramatic. Let p be the relative number of shortcuts in a small-world network (with respect to the overall number of con-

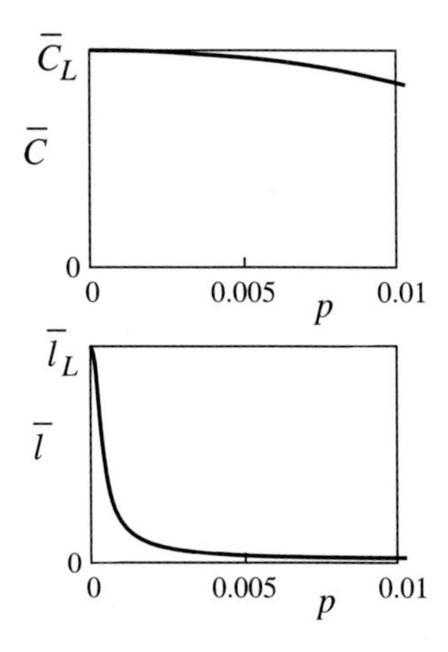

Fig. 3.4 A sketch of two typical dependences for small-world networks. Mean clustering $\overline{C}$ versus the relative number of shortcuts p and average internode distance $\overline{\ell}$ versus p, in a small-world network with a fixed finite number of nodes. $\overline{C}_L$ and $\overline{\ell}_L$ are the corresponding values for the mother lattice without shortcuts.

nections. Watts and Strogatz observed that even at a very low p, where the clustering is still nearly the same as in the mother lattice, the mean internode distance is tremendously reduced. Figure 3.4 shows schematically these typical dependences of clustering and the average distance between nodes on p in a finite small-world network. Thus even at small (but non-zero) p the network shows the small-world phenomenon.

One should stress that in these networks there is no sharp transition between two regimes: from large to small worlds. Rather, a smooth crossover is realized as the number of shortcuts grows. In particular, in the limit of the infinite number of nodes N, we obtain a 'large world' if the number of shortcuts N_s is finite, and a small world if the relative number of shortcuts p is finite. Let us estimate the average distance between nodes, $\overline{\ell}(p)$. We assume that $N \to \infty$ and p is finite, that is, the number $N_s \sim pN$ of shortcuts is large enough—the small-world regime. In this case the shortcuts determine the global architecture of the network, and the network resembles a classical random graph with N_s links and of the order of N_s 'supernodes'. These 'supernodes' are regions of the mother lattice, which contain neighbouring ends of shortcuts. To this graph, we can apply a classical formula (2.5), that is $\overline{\ell} \sim \ln N$, with two changes. In our estimation we ignore constant factors and for the sake of simplicity suppose that the network is based on a one-dimensional mother lattice—a chain. In the resulting network, the average distance between ends of different shortcuts is $\overline{d} \sim 1/p \sim N/N_s$. Then, (i) substitute N for N_s; (ii) take into account that the shortest paths in the network pass the mother lattice between ends of different shortcuts, so multiply the classical expression by $\overline{d}$. This gives the result:

$$\overline{\ell}(p) \sim \overline{d} \ln N_s \sim (1/p) \ln(Np)\,. \tag{3.1}$$

Thus the diameter of this small-world network is about $N_s / \ln(N_s) \gg 1$ times smaller than that of the mother lattice.

We can obtain relation (3.1) in a slightly different way. The point is that often, large small-world networks may be treated as locally tree-like. Look at Fig. 3.5 which shows a small-world net based on a simple chain. One can see that as $N \to \infty$, any finite environment of an arbitrary node is tree-like. The figure also shows this local structure. The reader can apply the arguments, which we used in Section 2.3, to this tree-like structure, estimate the mean branching, and rederive result (3.1).

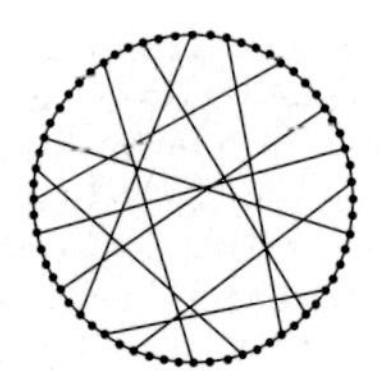

Fig. 3.5 This small-world network has locally a structure of a tree.

3.4 Equilibrium versus growing trees

Up to now we have discussed equilibrium random networks although most real networks are non-equilibrium. The difference between networks from these two classes is sometimes great. For demonstration purposes we compare here two important random trees—equilibrium and growing. These two models are selected to be as close to each other as possible the only difference being: one tree is equilibrium and the second—growing.

Let us first introduce recursive networks. A recursive network grows in the following way. Add a new node and attach it to a number of already existing nodes. Then repeat again, and again, and again. Figure 3.6 (a) explains this growth process (the bubble denotes an existing network). The nodes for attachment are selected by rules defined in specific models. Notice that in this growth, new links cannot emerge between already existing nodes. This is, of course, a great simplification, which makes these networks easy for analysis. Most growth models that we will discuss are of this type. Let us now suppose that in the recursive process, each new node has only one link, and the growth starts from a single node, see Fig. 3.6 (b). Then this network is a recursive tree (has no loops). Here the attachment is also a sequential birth of new nodes by existing ones. One should specify how a node for attachment is selected. The simplest rule is to choose it from among the existing nodes with equal probability, that is uniformly at random. This gives *a random recursive tree* as it is known in graph theory. This construction provides us with a maximally random tree with a given number of nodes under only one restriction: this tree is obtained by sequential addition of labelled nodes—the causality restriction. The random recursive tree is used as a reference model for growing networks. We will consider this important random graph in more detail in Section 7.1. Note that its degree distribution significantly differs from that of classical random graphs. For large recursive trees, the degree distribution approaches the exponential form: $P(q) = 2^{-q}$.

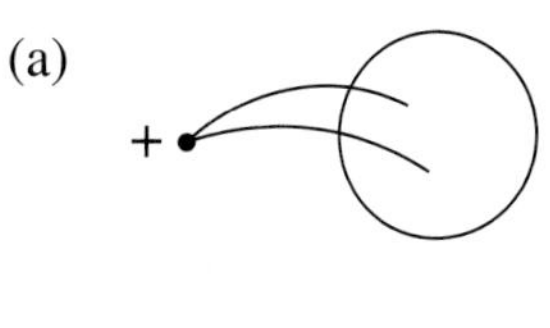

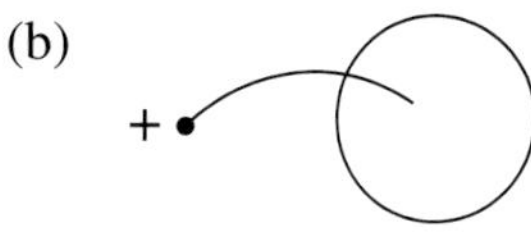

Fig. 3.6 How recursive networks grow. (a) The growth of a loopy recursive network. Each new node is attached to a few chosen existing nodes. Note that new connections are only between an added node and already existing nodes. (b) The growth of a recursive tree. (We assume that the growth starts from a single node.)

It is an easy exercise to find that the mean internode distance in the tree is $\overline{\ell}(N) \cong 2 \ln N$, if the number N of nodes is large.[3] Thus this random tree is a small world. Let us compare the random recursive tree with an equilibrium tree. By definition, this is a maximally random tree with a given number of nodes N. Compare with the definition of the Erdős–Rényi model. In other terms, this statistical ensemble contains all possible trees of N nodes taken with equal statistical weight. Figure 3.7 shows the members of both of these ensembles for a few small values of N.

[3] Compare the total lengths of the shortest paths between all pairs of nodes in this network at 'times' N and $N+1$. These are $\overline{\ell}(N)N(N-1)/2$ and $\overline{\ell}(N+1)(N+1)N/2$, respectively. The difference due to the new attached node is $1+[\overline{\ell}(N)+1](N-1)$. Together these three terms give an equation for $\overline{\ell}(N)$, which can easily be solved.

One can see that in the random recursive tree some configurations

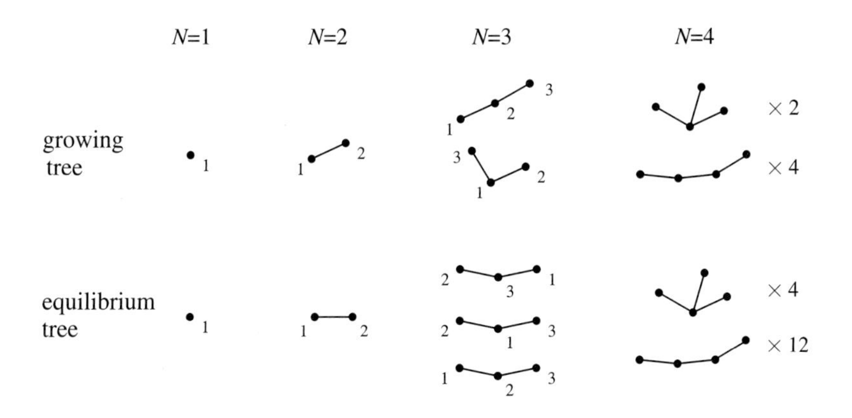

Fig. 3.7 The random recursive tree versus the equilibrium tree. The upper and lower rows shows the members of these two statistical ensembles. Each of them has a unit statistical weight. At $N=4$ we indicate the number of configurations which differ from each other only by labels.

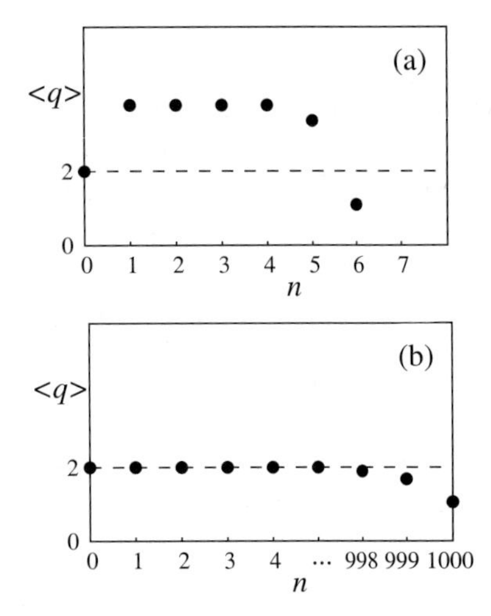

Fig. 3.8 The average degree of the n-th nearest neighbour of a randomly chosen node as a function of n for two networks: for a particular small world—a recursive tree (a), and for an equilibrium tree—a fractal (b).

are absent. For example, there is no configuration 2—3—1 (equivalent to 1—3—2). Already at $N=4$, the equilibrium tree has a larger mean internode distance than that of the growing one (compare weights of more and less compact configurations in the figure). As N increases, this difference becomes vast. One can show that the diameter of the equilibrium random tree scales as $\sqrt{N}$ at large N. Therefore it is not a small world but rather a random fractal with Hausdorff dimension 2. Interestingly, the degree distribution of this network is close to classical random graphs. In the former case it is simply factorial: $P(q) \propto 1/(q-1)!$. Nonetheless, one of these networks is a small world and the second—a 'large' one.

In Fig. 3.8, the readers can see how different are the neighbourhoods of a randomly chosen node in a random recursive tree (a) and in an equilibrium tree (b). The figure shows the dependences of the average degree of the n-th nearest neighbour of a node on n. Clearly, $\langle q\rangle(n=0)$ coincides with the average degree $\langle q\rangle$ of nodes in a network. This average degree approaches 2 in any infinite tree. A significant difference is for $n \geq 1$. In this range, for the small-world phenomenon the mean branching must be greater than 1. So in the recursive random trees $\langle q\rangle(n) > 2$. In fact, the dependence, schematically presented in Fig. 3.8 (a), is valid for any small world. Of course, $\langle q\rangle$ in loopy networks will be greater than 2. As for the equilibrium trees, the mean branching is close to 1 even in a very far neighbourhood of a node.

3.5 Giant connected component at birth is fractal

This 'fractal appearance' of the equilibrium trees has remarkable consequences even for loopy networks with the small-world property. The point is that we can find equilibrium trees within loopy small worlds. For an arbitrary connected graph, one can construct a tree which spans over the entire graph. This subgraph, the *spanning tree*, by definition, consists of all nodes of a given graph and of some of its links. Clearly, if a given graph is not tree, then it has a number of different spanning trees. In principle, this number may be very large. *A uniform (or random) spanning tree* of a given graph is a statistical ensemble whose members are all spanning trees of this graph, taken with equal probability. In simple terms, this is the maximally random spanning tree of a given graph. One may suppose that this random tree is equilibrium. Furthermore, it turns out that the uniform spanning trees of classical random graphs are fractals of Hausdorff dimension 2 [99]. In the preceding section we explained how to make a small world from a 'large one' by using long-range shortcuts. Contrastingly, the uniform spanning trees enable us to make a large world from a small one.

Even though the classical random graphs are loopy networks, they contain plenty of trees. Note that classical random graphs are only locally tree-like. So their finite components are (almost surely) trees, but a giant connected component is loopy—it has long loops. Recall our estimation of the diameter of a classical random graph in Section 2.3. We emphasize here that the result $\overline{\ell} \approx \ln N/\ln\langle q\rangle$ was obtained for an equilibrium, locally tree-like network which, however, has many long loops. We mentioned that this estimate is valid only sufficiently far from the birth point of a giant connected component. Suppose that we are approaching the birth point $\langle q\rangle = 1$ from above, and so a giant component increasingly resembles finite ones. So, it has fewer and fewer loops. In the limit $\langle q\rangle \to 1$, the giant component becomes an equilibrium tree. One can show that it has a fractal architecture, and its Hausdorff dimension equals 2, as is natural for random trees of this kind. Thus in the limit $\langle q\rangle \to 1$, the classical random graphs are fractals!

We can easily reach this fractal state by removing at random nodes or links in a graph, until a giant connected component disappears. In a large but still finite random graph, one can also make the following operation. Choose at random a node within a giant connected component. Then remove this node, and all of its neighbours closer than distance n from it. When n is sufficiently small, we will get a hole in the original giant connected component. With growing n, this hole increases, and at some critical value, it splits a giant component into a set of finite ones, see Fig. 3.9. Close to this point, the vanishing giant component again turns out to be a two-dimensional fractal [161]. In that sense, the remote part of a classical random graph with a removed 'centre' has a fractal structure.

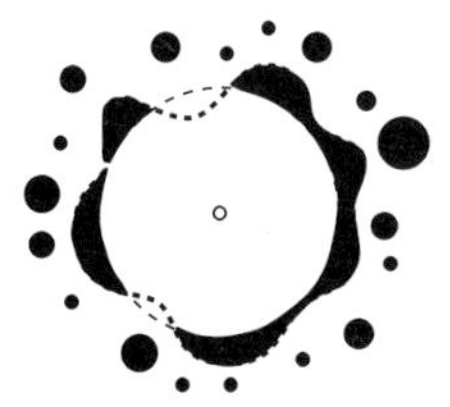

Fig. 3.9 The remnant of a random graph after the removal of a large environment of a node (open dot). A giant connected component on the verge of vanishing has a fractal architecture.

3.6 Dimensionality of a brush

We introduced small worlds as objects with an infinite fractal or Hausdorff dimension. Recall how the Hausdorff dimension was defined in terms of the size dependence of the mean internode distance: $\overline{\ell}(N) \sim N^{1/d}$, that is $d = \ln N/\ln\overline{\ell}$. One should note that this definition of the dimensionality of a system is not unique. Let us discuss an alternative way using a random walk on a network. A random walk here plays the role of a useful instrument which allows us to characterize the structure of a network. Consider a particle which at each time step, with equal probability, moves from a node to one of its nearest neighbours. After t steps of a random walk on a regular d-dimensional lattice, the particle will typically be found at a distance of the order of $\sqrt{t}$ from a starting point. So the 'area' of the region where the particle may be found at time t is about $t^{d/2}$. This is simply the number of nodes in the hyperball of radius $\sqrt{t}$. Then the probability that the particle will be found at the starting point after t steps is $p_0(t) \sim t^{-d/2}$. This can be used as another definition for the dimensionality. Remarkably, for some objects, this definition results in a dimension which differs from the Hausdorff one. So it is natural to introduce a special number, a *spectral dimension* d_s, defined by the relation $p_0(t) \sim t^{-d_s/2}$.[4] Actually, d_s shows that random walks in a given network are organized similarly to those on d_s-dimensional lattices.

[4] We will explain why it is called 'spectral' in future lectures.

A difference between spectral and Hausdorff dimensions may be illustrated by combs and brushes, see examples in Fig. 1.2 (c) and (d). Let a brush be based on a large d-dimensional lattice with long hairs—linear chains—growing from some of the nodes of a lattice. The Hausdorff dimension of this brush is $d+1$. As for its spectral dimension, it was found that $d_s = 1+d/2$ when d is not higher than 4, and $d_s = 3$ for $d \geq 4$ [107]. One can even use a network with the small-world feature and attach long linear chains to its nodes, and again, the spectral dimension is only 3.

From the Internet to cellular nets

4

A wide range of real-world networks from diverse areas have surprisingly similar architectures. In this lecture we discuss and compare structural properties of a few basic man-made and natural networks. Before starting, we emphasize the principal difference between two global networks—the Internet and the World Wide Web (WWW).

(i) The Internet is a global technological network of connected computers through which users can access data and programs from other computers. Links in this network can be wired or wireless.

(ii) The WWW is a global information network, an array of web documents (files of various formats), connected by hyperlinks. The hyperlinks are mutual references in web documents.

Even though most impressive, the WWW is only one of many Internet applications. Another one, for example, is email.

4.1 Levels of the Internet

The first distributed computer network ARPANET was constructed at the end of 1969.[1] Originally ARPANET linked only four nodes: the University of California at Los Angeles, Stanford Research Institute, the University of California at Santa Barbara, and the University of Utah. This pioneering, US national net afterwards grew into the Internet by interconnecting with other networks [53]. The key idea was to build the Internet as a federation of interconnected autonomous (independently managed), peer networks of very different types and architectures. The routing of data packets within the peer networks is maintained by their individual internal rules—protocols, while routing between these networks is performed by common internetwork routing protocols.[2] Thus, based on 'hierarchical routing', Internet technology was substantially formed by the middle of the 1980s. By 1991, the Internet included 700 000 host computers, and, in principle, approached the modern form.

The specific organization of the Internet as a network of numerous autonomous networks, without a central authority is apparently optimal and inevitable. 'The Internet is the first computational artifact that was not designed by one economic agent, but emerged from the distributed, uncoordinated, spontaneous interaction (and selfish pursuits) of many.

[1] The US Defense Advanced Research Projects Agency (DARPA) played a great role in the history of networking and initiated the 'Internetting' program in 1972. The organization of ARPA (1958, renamed DARPA in 1972) was initiated in response to the launch of Sputnik on October 4, 1957 by the Soviet Union.

[2] A protocol is an algorithm, a standard, a set of formal instructions and rules, see in more detail in Lecture 12.

[3] From the paper 'On a network creation game' by Fabrikant *et al.* (2003) [90].

Today's Internet consists of over 12,000 subnetworks ("autonomous systems"), of different sizes, engaged in various, and varying over time, degrees of competition and collaboration.'[3] This long quotation touches upon several key aspects of the evolving architecture of the Internet. Let us consider this multilayer architecture in more detail. The Internet includes hosts (computers of users), servers (computers or programs providing a network service), and routers, arranging the traffic. The total number of hosts (including handheld devices) in the Internet was about 570 million in July 2008 and will probably reach 3 billion (3 000 000 000) by 2011 [1]. Figure 4.1 shows schematically the multilayer architecture of the Internet. The complete structure of the Internet including all host computers has never been investigated. Routers with their undirected interconnections form the router level network in the Internet. The autonomously administered subnetworks (autonomous systems, AS) in this network are the nodes of the second, AS-level graph. Routing between autonomous systems is maintained by the common Border Gate Protocol. (A gateway is a system (software or a device) that joins two networks together.) In the AS graph, two autonomous systems are connected by a single undirected link if at least two of their routers are directly connected. Because of the exponentially rapid growth, it is even hard to estimate the present sizes of these two networks. Very approximately, the Internet contained 40–70 thousand autonomous systems in 2009. The number of routers was higher, say, by 15–25 times.

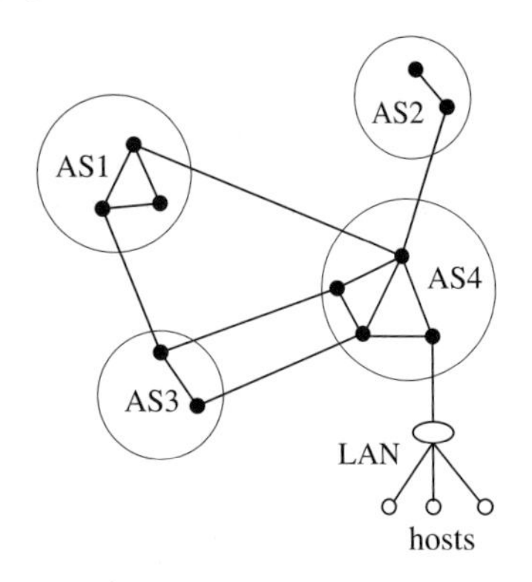

Fig. 4.1 Multilayer organization of the Internet. Open dots show host computers of users, LAN is a local area network, filled dots are routers, and ASs are autonomous systems.

The routers in the Internet have geographical locations. It turns out, however, that usually it is hard to find their precise coordinates. As for the autonomous systems, some of them contain routers distributed all over the world, and any geographical mapping is impossible. This is why the majority of studies have to ignore the geographical factor in Internet architecture. The statistics of connections in the Internet remained unstudied until the end of 1990s. In 1999, three scientists, Faloutsos, Faloutsos, and Faloutsos, analysed the partial maps of the AS and router networks in the Internet and discovered that both of them had not classical, but scale-free architectures. They found that the degree distributions of these networks could be approximately fitted by power laws [91]. The networks that they investigated were small, so in Fig. 4.2, we reproduce empirical cumulative degree distributions obtained later for larger AS and router maps by Pastor-Satorras, Vázquez, and Vespignani [148,177].[4] Both these empirical cumulative distributions were fitted by power-law dependencies with exponent 1.1, and therefore the degree distribution exponent $\gamma = 2.1$.

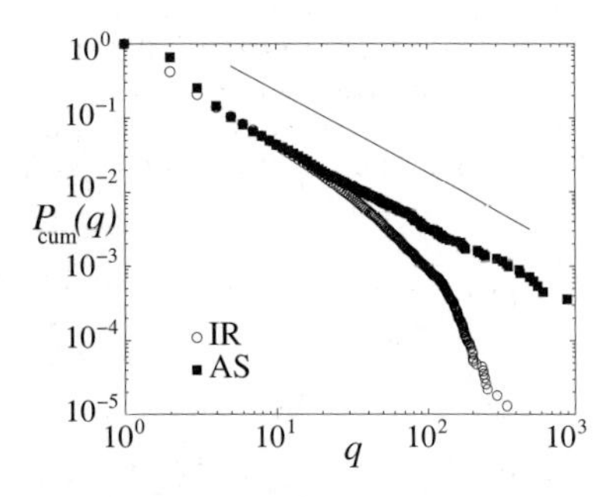

Fig. 4.2 The cumulative degree distributions of the two networks: AS level map measured in October/November 1999 and the Internet Router (IR) map of May 2001. Adapted from the paper of Vázquez, Pastor-Satorras, and Vespignani [177].

[4] The cumulative degree distribution is defined as $P_{\text{cum}}(q) = \sum_{q' \geq q} P(q')$. If $P(q) \sim q^{-\gamma}$, then $P_{\text{cum}}(q) \sim q^{1-\gamma}$. Researchers use these distributions to diminish inevitably strong fluctuations in the region of large degrees. Another method to smooth fluctuations is binning—accumulation or averaging data within some degree intervals.

The empirical distributions in Fig. 4.2 are very typical [10]. Note that the 'quality' of these 'power laws' is rather poor, especially for the router network, which is also a very typical situation. The AS network consisted of about 11 000 autonomous systems with, on average, 4.2 connections for each AS. The network of routers had about 230 000 nodes and the router's mean degree $\langle q \rangle \approx 2.8$. For smaller networks and higher values of exponent γ, empirical curves usually resemble even less a power law. Importantly, the Internet maps which empirical researchers analyse, are very approximate and always incomplete. These maps were obtained by sending packet probes over the network from one or a few sources. Unfortunately, this technique may seriously distort the empirical degree distributions. Nodes with few connections have a higher chance of escaping the probe, and the measured degree distributions are more skewed than the real ones [151, 56]. Nonetheless, despite all these difficulties and restrictions, the empirical data for the Internet show remarkably skewed degree distributions. The biggest autonomous systems have a few thousand connections to other ASs.

Both the networks have small diameters. The average internode distance in the AS network is only 3–4 hops. In the router network this distance was found to be about 9. These were data obtained in 1999–2001 for Internet maps. The Internet grows exponentially with time, and both the AS and router networks evolve rapidly. It was theoretically suggested that these and many other networks become more densely connected with time. In other words, the growth is 'accelerated' in the sense that the number of links in a network grows more rapidly than the number of nodes [78]. Empirical observations have confirmed that this is indeed the case [120]. The growing average degree results in increasing clustering. Furthermore this densification leads to shrinking diameters of the networks. The networks exponentially grow but their diameters are constant or even decrease with time.[5]

[5] Interestingly, with time, many routers and autonomous systems disappear.

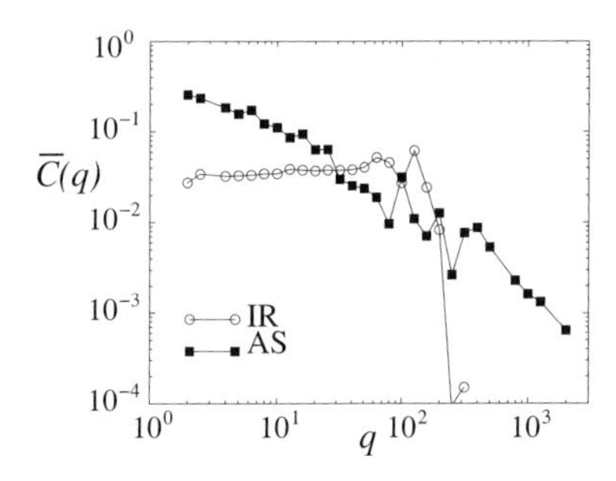

Fig. 4.3 The mean clustering of a node versus the degree of this node for the same two networks as above. Adapted from the paper of Vázquez, Pastor-Satorras, and Vespignani [177].

One can show that the heavy-tailed degree distributions make strong clustering inevitable. The measured values were $\overline{C} \approx 0.03$ and $\overline{C} \approx 0.3$ for mean clustering in the router and AS networks, respectively [177]. Moreover, in contrast to classical random graphs, the clustering of a node in the router and AS networks was observed to depend strongly on the degree of this node, see Fig. 4.3. We explained that in this situation, the clustering coefficient C can essentially differ from $\overline{C}$. In reality, C in these networks is much smaller than $\overline{C}$.

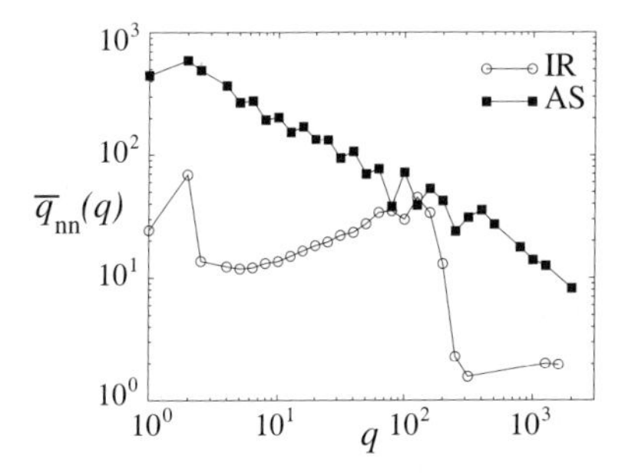

Fig. 4.4 The average degree of the nearest neighbour of a node of degree q for the same two networks as above. Adapted from the paper of Vázquez, Pastor-Satorras, and Vespignani [177].

Pastor-Satorras, Vázquez, and Vespignani also measured a quantity, which became one of the standard structural characteristics of complex networks. They investigated the average degree $\overline{q}_{\text{nn}}(q)$ of the nearest neighbour of a node with q connections. This quantity is independent of q in the uniformly random, that is, uncorrelated networks; see the next lecture. In these uncorrelated networks, nodes 'know' nothing about the degrees of their neighbours. In the Internet, this is certainly not the case. Figure 4.4 demonstrates the sharp difference between Internet networks and uniform ones. Therefore, it is not only the skewed degree distribution that distinguishes these complex networks from classical random graphs. Subsequent studies have shown that the Internet is not an exception. On the contrary, very few real-world networks have no structural correlations.

4.2 The WWW

In essence, the WWW is simply a system for automated retrieving of information in the form of electronic documents (files of various formats). The original idea was to organize links between documents in a way, convenient for users. In the WWW, the linking is based on *hypertext.* The hypertext contains highlighted parts which cover the link to other documents. Clicking on highlighted text causes the fulfilment of the underlying link to the corresponding document and downloads this file to the user's computer. In 1989, Tim Berners-Lee proposed a hypertext system for CERN, the European Laboratory for Particle Physics in Geneva, Switzerland [25, 65]. In 1990, he wrote a program 'WorldWideWeb', which is a web browser editor, 'a program which provides access to the hypertext world', and the first web page was placed on the first web server in CERN. Hypertext documents in the WWW are written by using special programming language, Hypertext Markup Language (HTML) (look at any web page by using the option *View the Page Source* of your browser). The functioning of the WWW is based on the Hypertext Transfer Protocol (HTTP), which enables the flow and processing of web documents and requests in the Web. Thus, in 1990, the four required components—(i) the protocol of the Web, HTTP, (ii) the language of the Web, HTML (which are the two main standards of the Web), (iii) a web browser, and (iv) web servers—were created. The WWW was born.

On the 25 July 2008 the official Google blog announced: the Google index 'hit a milestone: 1 trillion (as in 1 000 000 000 000) unique URLs (Uniform Resource Locator) on the web at once!'[6] This number—a trillion pages in the Google index, however impressive, does not allow us to estimate the size of the exponentially growing WWW. The difficulty is that only a very small fraction of pages are accessible by search engines. These public pages, sufficiently static to be scanned, form the so-called *Surface Web.* The huge part of the WWW, (*the Deep Web*), are electronic documents in databases and archives with restricted public access and, also, rapidly varying pages (time tables, web calendars, and so on) with all their hyperlinks. A clear border between the Surface and Deep Webs is absent, and it is barely possible to reliably evaluate their sizes. What empirical researchers can study are sufficiently large pieces of the WWW. The analysis of these parts allows us to understand the architecture of the WWW.

[6] See http://googleblog.blogspot.com/2008/07/we-knew-web-was-big.html.

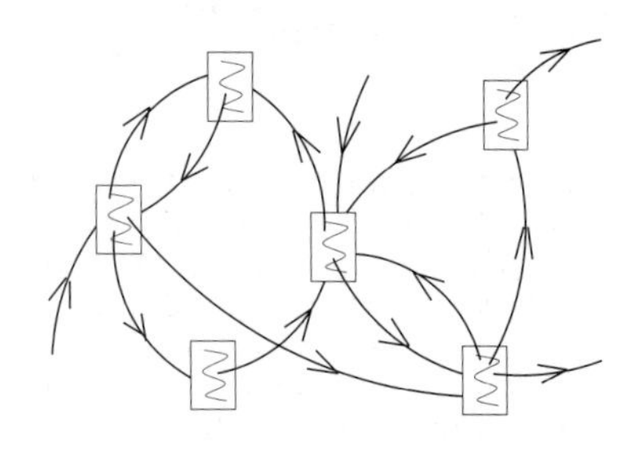

Fig. 4.5 Connections in the WWW. Notice reciprocal hyperlinks.

Figure 4.5 shows very schematically how hyperlinks connect web documents. Compare this schematic view of directed connections in the WWW with a more simple picture of connections in the network of citations in scientific papers, see Fig. 1.7. Notice reciprocal hyperlinks in the Web graph. Two Web documents can cite each other while scientific papers cannot. It turned out that the number of reciprocal hyperlinks in the WWW is surprisingly large, up to 60% of all connections [85, 138]. The abundance of reciprocal hyperlinks in the WWW becomes more clear if we look at the scheme of connections of a typical home page, see Fig. 4.6. These figures also show longer loops. The directedness of the hyperlinks make makes these loops more diverse than in undirected networks. For example, there are six different loops of length 3 if we also take into account reciprocal links. In principle, the definition of clustering must be modified to account for this diversity. Usually, however, the clustering of the WWW network is measured ignoring the directedness of connections, which typically gives $C \approx 0.1$ and $\overline{C} \approx 0.3$.[7]

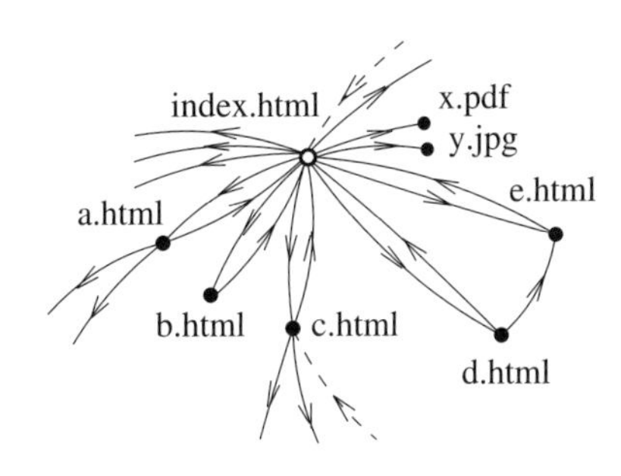

Fig. 4.6 Structure of a typical home page. The dashed arrows show hyperlinks coming from external web documents.

The directedness of hyperlinks determines a more rich and interesting organization of finite and giant components in the WWW than in undirected networks. According to a standard definition, a giant component obtained ignoring the directedness of connections is *a giant weakly connected component*. For a general directed network, this giant component is organized as is shown in Fig. 4.7. There is *a giant strongly connected component*, which consists of the nodes mutually reachable by directed paths. The nodes of *a giant out-component* are reachable from the strongly connected component by directed paths. *A giant in-component* contains all the nodes from which the strongly connected component is reachable. By this definition, the giant strongly connected component is the intersection of the giant in- and out-components. The remaining part of the giant weakly connected component is the mess of so-called tendrils. Apparently, the presence of a giant strongly connected component is vitally important for the function of the WWW. In 1999, Broder and coauthors measured the sizes of these components using a map of a sufficiently large part of the WWW (about 200 million pages) [44]. They found that the giant strongly connected component contained about 30% of the pages in the weakly connected component, while the tendrils contain about 25%. The average length of the shortest direct path between two web pages in these measurements was about 16. Remarkably, the maximum separation of nodes observed in this network was very large, namely, about 1000 clicks!

[7] These values are for the same nd.edu domain of the WWW, which we discussed in the previous lecture [5]. Surprisingly, taking into account the hyperlink directedness leads to even smaller numbers of triangles than one could expect [31].

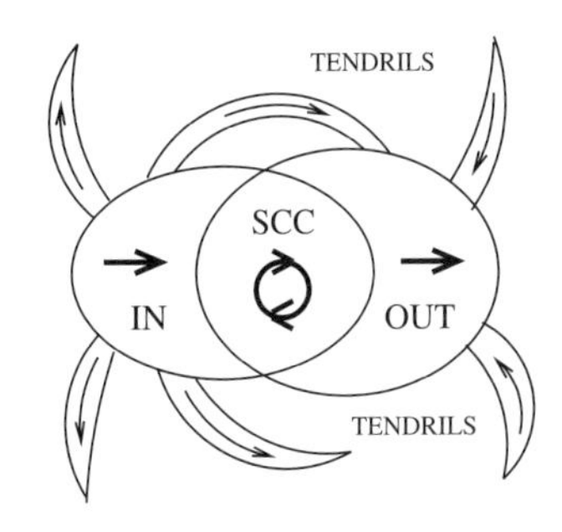

Fig. 4.7 Organization of a giant weakly connected component in a directed network. SCC, IN, and OUT are the giant strongly connected component and the giant in- and out-components. For more detail, see [80]. Compare with Broder *et al.* [44].

Similarly to the Internet, the density of connections in the WWW grows with time. Broder and coauthors first made their measurements in May 1999 and found that the average in- and out-degrees of a node are equal at 7.22. When they repeated the measurements in October 1999, the average in- and out-degrees were already 7.85.

The scale-free in- and out-degree distributions of the WWW, $P_{\rm i}(q_{\rm i}) \sim q_{\rm i}^{-\gamma_{\rm i}}$ and $P_{\rm o}(q_{\rm o}) \sim q_{\rm o}^{-\gamma_{\rm o}}$, respectively, were observed in 1999 by Albert, Jeong, and Barabási, who studied a relatively small nd.edu domain. Broder and coauthors (2000) obtained the degree distributions for a much larger network [44]. They fitted the empirical in- and out-degree distributions by power-law dependencies, with exponents $\gamma_{\rm i} = 2.1$ and $\gamma_{\rm o} = 2.7$, respectively. The power law for $P_{\rm i}(q_{\rm i})$ was observed in a wide range of in-degrees (about three orders of magnitude), and so the obtained value 2.1 is reliable. This is not the case for the out-degree distribution. The range of out-degrees in the WWW is much narrower than that of in-degrees, so there is even a doubt that a power law for out-degrees exists at all [71].[8] In any case, in respect of at least the in-degree distribution, the WWW is a scale-free network.

[8] It is impossible to put on a page, say, one million links to other web documents.

4.3 Cellular networks

The Internet and WWW networks are sufficiently large to allow reliable statistical analysis of their architectures. Even on the AS level, the Internet network has many thousand nodes. In this section we will touch upon a few really small cellular and genetic networks, whose specific structures could be revealed despite their tiny size.[9] In 2000, Jeong, Tombor, Albert, Oltvai, and Barabási made a thorough study of the networks of the metabolic reactions of 43 simple organisms belonging to different domains of life [106]. All these networks were very small, from 200 to 800 nodes. The network of metabolic reactions, in essence, is a typical chemical reaction graph. Its nodes are molecular compounds participating in various metabolic reactions as educts or products. The same compound can be educt in one reaction and product in another. If two compounds participate in some reaction as an educt and a product, connect them by a directed link going from educt to product. Thus Jeong and coauthors had 43 sparse directed graphs to analyse. The remarkable conclusion was that the in- and out-degree distributions of all 43 networks are approximately scale-free with exponent equal to 2.2. The definite conclusion for such small networks was possible due to the low value, 2.2, of the degree distribution exponent. The lower this exponent, the wider the range of degrees to observe a power law. This scale-free architecture has a direct consequence on the distributions of metabolic reaction fluxes in these networks. These distributions were also found to be skewed [86, 8].

[9] For detailed discussion of complex network concepts in cellular biology, see the review of Barabási and Oltvai [15].

The second example of a cellular network, which we touch upon, is the network of physical protein interactions of the yeast *Saccharomyces cerevisiae* [105]. This network of 1870 nodes and 2240 undirected links is bigger than the metabolic networks that we have discussed in this section, nonetheless, the statistical data are less conclusive. The nodes of this network are different proteins and the undirected links are physical interactions (direct contacts) between them. It was impossible to fit the empirical degree distribution by a power-law function. Instead, the distribution was fitted by a power law with an exponential cut-off. The power-law exponent can be roughly estimated as 2.5, which, we believe, is already too large to reliably observe scale-free distributions in networks of this size. Still, the general architecture of this network is rather similar to that of, say, the AS-level Internet network. In particular, the empirical dependencies $\overline{C}(q)$ and $\overline{q}_{\rm nn}(q)$ look similar to those in Figs. 4.3 and 4.4 [122, 123].

In both these examples, the networks are clearly defined, and a graph for empirical study can be obtained in a quite strict way. For many other systems, an underlying network structure is not that obvious. Using genetic networks as an illustration [156], let us demonstrate how the network structure can be unveiled.

Without going into detail, *genome* is a large set of interacting genes which encode the genetic information of an organism. A given living organism is characterized by a set of features, the so-called gene expressions. The expression of each gene is quantitatively described by its level—*an expression level.* The term 'interaction between genes' is used here in the sense that genes function not independently, but in cooperation: the expression of a gene depends on other genes. Consequently, expression levels are not independent. These correlations between expression levels of genes are treated as the result of the cooperative function of genes, that is of their 'interaction'.

The procedure for obtaining correlations between gene expression levels is routine in genetic research. Suppose that the expression levels of N genes in a genome are e_i, $i = 1, \ldots, N$. The original ('wild-type') cell culture is 'perturbed' M times. For example, the culture is exposed to radiation and mutant strains with distorted genomes are produced. The full set of gene expression levels, $\{e_i^{(s)}\}$, $s = 1, \ldots, M$, of these mutant strains is measured (see Fig. 4.8). The correlation c_{ij} between the expression levels of two genes, i and j, may be easily obtained by averaging over all the mutant strains:[10]

$$c_{ij} = \langle e_i e_j \rangle - \langle e_i \rangle \langle e_j \rangle \,, \tag{4.1}$$

where the average $\langle\ \rangle = M^{-1} \sum_s$. Correlations between genes i and j are absent when $c_{ij} = 0$. The resulting numbers c_{ij} are the elements of a large $N \times N$ matrix which shows how different genes interact with each other. However, most of the matrix elements are small and inessential, and this matrix, in its original form, provides superfluous information. For unveiling the basic structure of gene interactions, one must take into account only the important matrix elements. So, genes i and j are believed to be interacting to (be connected) if only the element c_{ij} is greater than some threshold value. This value is chosen such that the resulting network is sufficiently sparse, and the structure of essential pairwise interactions is clearly visible.

The same approach may be applied to many other situations. In general, network constructions of such a kind use two basic restrictions:

(i) Take into account only 'important' pairwise interactions or correlations between 'nodes'.

(ii) Ignore the difference between the magnitudes of these interactions.

In principle, one can present the complete information in the form of a *weighted network*, where each link has its real number value c_{ik}, taken, for example, from eqn (4.1). Usually, however, these weighted networks are too informative for a direct analysis.

Expression levels
Genes
Strain 1
Strain M

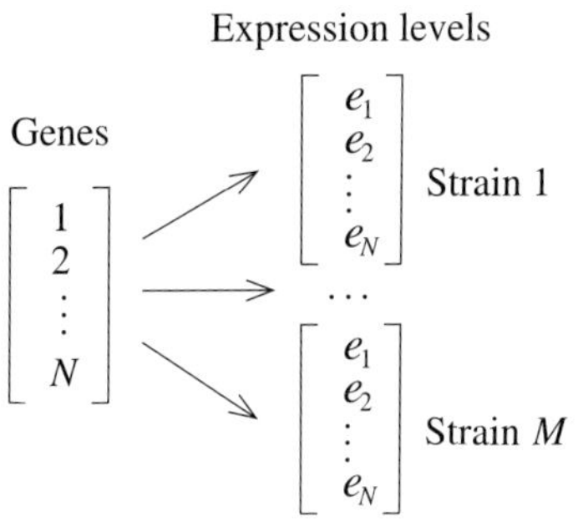

Fig. 4.8 Genes in a genome (N genes) are expressed in terms of characteristic features—traits—of the organism (N gene expression levels). These features are not independent. Exposure to radiation (M times) leads to mutant strains with M sets of gene expression levels. The statistical analysis of these sets allows one to find correlations between different gene expressions and to unveil the cooperative function of the corresponding genes.

[10] This is a simplified form of the expression. Moreover, instead of the expression levels e_i, the logarithms of the ratios of expression levels of mutant strains and those of the wild-type cell culture are actually used.

4.4 Co-occurrence networks

In many systems allowing network representation, groups of elements co-occur in various associations. That is, a given set of elements each have something in common, or they participate together in some action, or, they present in the same list, and so on. If these associations are pairwise, then we have a simple network (nodes connected by ordinary single links). On the other hand, if the associations are more complicated, then we arrive at multi-partite graphs and hypergraphs. Let us touch upon these more interesting constructions.

The standard example of these networks is the bipartite network of directors sitting on many boards. We will consider this in the next lecture. Here we mention another typical example, namely the human disease network constructed by Goh and coauthors in 2007 [96]. The network was based on a long list of human genetic disorders and all known disease genes in the human genome. The disorders and disease genes are nodes of two types. If mutations in a gene are involved in some disorder, connect these two nodes by an undirected link. Since the same gene may be implicated in various disorders, the network has a rich structure of connections. The researchers hope that analysing the global organization and statistics of connections in this bipartite network will allow them to find as yet unknown relations between genetic disorders and disease genes. The point is that the complete system of disorders and disease genes is too large to uncover all the relations of this kind without using network representation.

In principle, the human disease network is a quite typical bipartite graph. Zlatić, Ghoshal, and Caldarelli (2009) studied more exotic networks [186] based on file-sharing databases Flickr and CiteULike. In these databases, users upload photos (Flickr) or put links to scientific papers (CiteULike) with short text descriptions. All registered users can supply these photos and papers by keywords—descriptive tags. The resulting network has three sorts of nodes: (i) users, (ii) photos (papers), and (iii) tags. When a user uploads a photo with a tag or assigns a tag to a photo uploaded by another user, he creates a new hyperedge interconnecting three nodes: this user, the photo (paper), and the tag; see Fig. 4.9. So these networks are tripartite hypergraphs, in which every hyperedge interconnects three nodes of different types. In these tagged social networks, there are three 'hyperedge distributions', for each kind of node. The researchers found that these distributions differ significantly from each other. The degree distribution for photos (papers) decays most rapidly of the three, the distribution for users decays the most slowly (apparently, there exist 'crazy taggers'), and the degree distribution for tags demonstrates an intermediate rate of decay. Thus, interestingly, there are no really 'superpopular' photos and papers either in Flickr or in CiteULike. Instead, there are plenty of people interested in tagging!

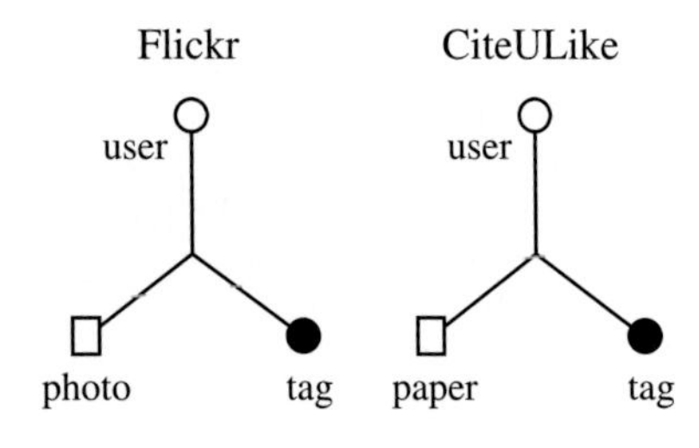

Fig. 4.9 Hyperedges in Flickr and CiteULike.

Uncorrelated networks

5

Most real-world networks are very far from the classical random graph models. In this lecture we show that these models can be greatly improved.

5.1 The configuration model

By the late 1970s, the theory of classical random graphs was well developed, and mathematicians started to search for more general network constructions. In 1978, Edward A. Bender and E. Rodney Canfield published a paper entitled 'The asymptotic number of labelled graphs with given degree sequences' [22], where they described random networks with essentially richer architectures than the Erdős–Rényi graph. Béla Bollobás strictly formulated this generalization of the Erdős–Rényi model in his paper 'A probabilistic proof of an asymptotic formula for the number of labelled random graphs' (1980) and named it the *configuration model* [38]. This generalization turned out to be a major step toward real networks in the post-Erdős epoch.

Before introducing the configuration model, we have to recall the idea of classical random graphs. A classical random graph is the maximally random network that is possible for a given mean degree of a node, $\langle q \rangle$. The Erdős–Rényi model is one of the two versions of classical random graphs: a maximally random network under two restrictions: (i) the total number of nodes N is fixed and (ii) the total number of links is fixed. These constraints result in the Poisson form of the degree distribution which differs from the degree distributions of real-world networks. To make a step towards real networks, one should be able to construct a network with, at least, a real degree distribution $P(q)$, and not only a real mean degree. The idea was to build the maximally random network for a given degree distribution. The configuration model provides a way (more precisely, one of the ways) to achieve this goal by directly generalizing the Erdős–Rényi construction. In graph theory, the term 'sequence of degrees' usually means the set of numbers $N(q)$ of nodes of degree q in a graph, $\sum_q N(q) = N$ [128]. Let this sequence be given. The configuration model is a statistical ensemble whose members are all possible labelled graphs, each with the same given sequence of degrees. All these members are realized with equal probability. We have explained that this corresponds to the the maximum possible randomness—uniform randomness.

The same can be done in the following way. Consider a set of N

nodes: $N(1)$ nodes of degree 1, $N(2)$ nodes of degree 2, and so on; supply each of the nodes by q stubs of links; choose stubs in pairs at random and join each pair together into a link (see Fig. 5.1).[1] As a result we have a network with a given sequence of degrees, but otherwise random. Clearly, if every node has the same number of connections, then this network is reduced to a random regular graph. On the other hand, if the sequence of degrees is drawn from a Poisson distribution, then (in the infinite network limit) we get a classical random graph.

[1] Note that the term 'at random' without further specification usually means 'uniformly at random'. To build a particular realization of this network, make the following: (i) create the full set of nodes with a given sequence of stub bunches; (ii) from all these stubs, choose at random a pair of stubs and join them together; (iii) from the rest of the stubs, choose at random a pair of stubs and join them together, and so on until no stubs remain.

The heterogeneity of these networks is completely determined by degree distributions; correlations are absent in contrast to real-world networks. Fortunately, the majority of phenomena in complex networks can be explained qualitatively based only on the form of degree distribution, without accounting for correlations.

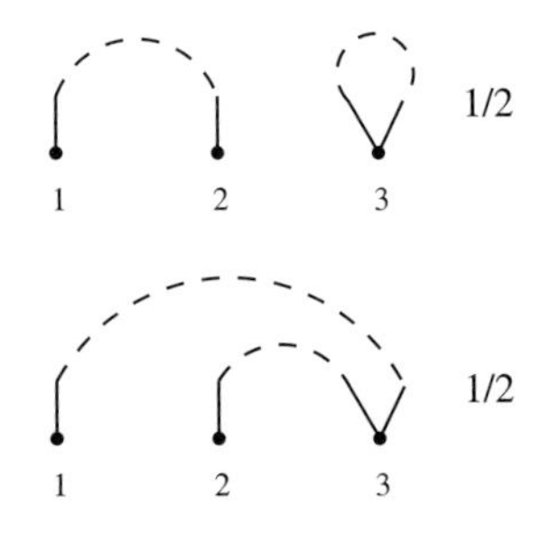

Fig. 5.1 A random network ensemble in the configuration model. In this example, $N(1)=2$ and $N(2)=1$. The two members of the ensemble have equal probability of realization, 1/2.

5.2 Hidden variables

The second way to build an uncorrelated network with an arbitrary degree distribution was found quite recently, in 2001–2002. This construction [97, 55, 48] directly generalizes the Gilbert (that is the $G_{N,p}$) model and is essentially based on the notion of 'hidden variables'. The general idea of the algorithm used is very simple. (i) To each of the nodes, $i = 1, 2, \ldots, N$, ascribe a number—*a hidden variable*, d_i. (ii) Then connect nodes in pairs with probabilities depending on the hidden variables of these nodes, so that the probability that there is a link between nodes i and j is $p_{ij} = f(d_i, d_j)$. The architecture of the resulting random network is determined by the statistics of the hidden variables and the form of a given function $f(d, d')$. In particular, if $p_{ij} = p$ is a constant, we arrive at the Gilbert model.

Suppose we aim to get an uncorrelated network with a degree distribution $P(q)$. Then use the 'desired degrees' as the hidden variables. Namely, (i) ascribe 'desired degrees' d_i—random numbers drawn from the distribution $P(d)$—to N nodes, and (ii) connect nodes in pairs (i, j) with probabilities p_{ij} proportional to the products $d_i d_j$. Normalization gives $p_{ij} = d_i d_j / N\langle d\rangle$ if we assume that the network is sparse. One can prove that the degree distribution of the resulting network approaches the desired form $P(q)$ at large degrees. One can also show that this network is uncorrelated. Chung and Lu (2002) called this construction 'the random graphs with a given desired sequence of degrees' [55]. This random network and the configuration model approach each other when the networks are infinite. However, networks built by using hidden variables are more convenient for practical purposes than the configuration model. In modern numerical experiments (simulations), researchers usually use this easy algorithm or its variations to build uncorrelated networks, and not the configuration model. Furthermore, these networks are easily treatable analytically.

Maybe, the reader has already noticed a serious problem with this construction. If the product $d_i d_j$ exceeds $N\langle d\rangle$, the probability p_{ij}, defined as above, becomes greater than 1, which is pure nonsense. This problem certainly arises for slowly decaying distributions $P(d)$. A simple patch to remedy this flaw was proposed in the very first work introducing a construction with hidden variables (2001) [97]. Its authors—Goh, Kahng, and Kim—proposed to restrict the probability p_{ij} from above by choosing this probability in the form: $p_{ij} \propto 1 - \exp(-d_i d_j / N\langle d \rangle)$.[2] The resulting construction is called the *static model*. The flaw has been fixed but, in return, another problem has emerged. For slowly decaying distributions $P(d)$, a network built with this p_{ij} appears correlated. Without going into detail, these networks are uncorrelated only if their degree distributions decay sufficiently rapidly. For more detail, see Lecture 8.

[2] This is, of course, only one of possible forms.

5.3 Neighbour degree distribution

In Section 2.3, we have shown that if the degree distribution of an arbitrary network is $P(q)$, then the degree distribution of any of the end nodes of a randomly chosen link is equal to $qP(q)/\langle q \rangle$. We stress that this is the case for any network. Let us introduce a joint distribution $P(q, q')$ of the degrees q and q' of the end nodes of a randomly chosen link. In uncorrelated networks, these degrees, q and q', are independent. So for these networks, a joint degree–degree distribution takes the following factorised form:

$$P(q, q') = \frac{qP(q)}{\langle q \rangle} \frac{q'P(q')}{\langle q \rangle}. \tag{5.1}$$

This also means that the branching of a link does not depend on the degree of its second end. The mean degree of an end of a randomly chosen link is $\langle q^2 \rangle / \langle q \rangle$. This is also the mean degree of a randomly chosen node. The mean branching is smaller by 1, and so, as we already know, $\bar{b} = \langle q^2 \rangle / \langle q \rangle - 1$. Note that this is also the ratio of z_2 (the average number of the second nearest neighbours of a node) and z_1 (the average number of the nearest neighbours of a node, $\langle q \rangle$).

It is important that almost always $\langle q^2 \rangle / \langle q \rangle$ is greater than the mean degree $\langle q \rangle$ in the network.[3] If a degree distribution decays slowly, the difference may be great. For various properties of networks, it is the organization of connections of the nearest neighbours of a node that matters. In respect of these properties, a network seems more dense than it really is. This basic observation explains a great number of phenomena in complex networks. We will exploit this extensively.

[3] Check that the equality $\langle q^2 \rangle / \langle q \rangle = \langle q \rangle$ takes place only if a network has no nodes other than the bare ones or the dead ends.

5.4 Loops in uncorrelated networks

The first thing we should do is to find whether these networks are loopy or not. Actually, we could expect that they have very few loops by analogy with classical random graphs, but we must check. Let us, to be concrete, find the clustering coefficient C in the configuration model. Let a network be large, $N \gg 1$. We should find the probability that two nearest neighbours of a randomly chosen node have at least one link between them (multiple connections are allowed). Suppose that these two nodes, i and j, have q_i and q_j connections, respectively, see Fig. 5.2. Then we have $q_i - 1$ stubs at the first node and $q_j - 1$ stubs at the second, and also nearly $N\langle q\rangle$ stubs at the other nodes. We use the fact that in the configuration model these stubs are connected in pairs at random. Then a stub at node i is connected to one of the stubs at node j with probability $(q_j - 1)/(N\langle q\rangle)$. So for all $q_i - 1$ stubs together we get the total probability $(q_i - 1)(q_j - 1)/(N\langle q\rangle)$. Now average this expression over degrees q_i and q_j, taking into account that these degrees are distributed as $qP(q)/\langle q\rangle$. This readily gives the clustering coefficient of an uncorrelated network [133]:

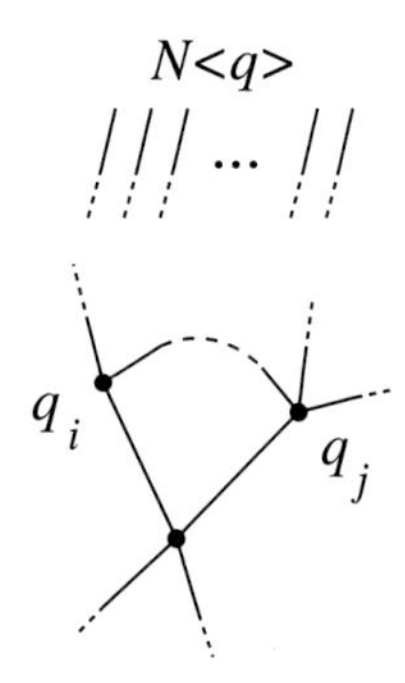

Fig. 5.2 Calculating the clustering coefficient in the configuration model. All the stubs in a network are connected in pairs at random, with equal probability.

$$C = \overline{C} = \frac{1}{N\langle q\rangle}\left(\frac{\langle q^2\rangle - \langle q\rangle}{\langle q\rangle}\right)^2 = \frac{\overline{b}^2}{N\langle q\rangle}. \tag{5.2}$$

For a Poisson degree distribution, this result is reduced to the expression $C = \langle q\rangle/N$ for classical random graphs.

At first sight, formula (5.2) does not look radically different from that for classical random graphs. The clustering vanishes in both cases as N approaches infinity. In other words, the clustering is only a finite-size effect in these models. Nonetheless, formally substituting empirical data ($\langle q^2\rangle$, $\langle q\rangle$, and N) for real networks into eqn (5.2) usually provides far more reasonable values than the classical random graph formula. Heavy tails in degree distributions of real networks lead to large $\langle q^2\rangle$, and this in turn results in a sufficiently high value of the calculated clustering coefficient at finite N. Typically, the classical formula underestimates C by several orders of magnitude—three, four, five orders, while with eqn (5.2) we often underestimate it by 'only' several times. Roughly speaking, the configuration model provides the smallest clustering that is possible in a random network of a given size with a given degree distribution, and ignores other details. Thus, clustering of many real-world networks turns out to be not that far from these 'minimum possible values'. To explain real values and size-independent contribution to the clustering coefficient, we should go beyond the configuration model and its variations.

Similarly, one can find the number $\mathcal{N}_L$ of loops of length L in an uncorrelated network. For sufficiently short (e.g., finite) loops, $\mathcal{N}_L \sim \overline{b}^L/(2L)$ [30, 32]. On the other hand, the number of loops longer than the diameter of a network is extremely large, $\ln \mathcal{N}_L \propto N$. In this respect, the situation is very similar to that for classical random graphs—few short loops and many long loops, which corresponds to a locally tree-like architecture.[4] This sea of long loops is a necessary feature of uncorrelated networks and, generally, of constructions of this kind. Thanks to the long loops, these networks have no boundaries, no centres. Their nodes can be distinguished only by their degrees, otherwise they are 'statistically equivalent', as physicists often say. We will show that this absence of boundaries is critically important for cooperative effects in complex networks.

[4] More rigorously, if the second moment of a degree distribution diverges, then the tree-like character disappears. This takes place in infinite networks with scale-free degree distributions if exponent γ is less than or equal to 3. Surprisingly, even in this difficult situation, the tree ansatz sometimes works.

5.5 Statistics of shortest paths

Using the tree ansatz we readily arrive at the asymptotic formula (large N) for the mean internode distance (and diameter) of an uncorrelated network:

$$\overline{\ell} \cong \frac{\ln N}{\ln \overline{b}} \,. \tag{5.3}$$

We have obtained this relation discussing classical random graphs. This result, typical for small worlds, is valid only if the second moment of a degree distribution is finite in the infinite network limit.[5] Otherwise, $\overline{\ell}$ can grow even slower than $\ln N$—*ultra-small worlds.* The resulting form $\overline{\ell}(N)$ depends on how $\langle q^2 \rangle$ approaches infinity with growing N. It is impossible to show a general formula, since this approach differs between different versions of uncorrelated network models. In particular, it may be $\overline{\ell} \sim \ln N / \ln\ln N$ for networks with exponent $\gamma = 3$ and $\overline{\ell} \sim \ln\ln N$ if exponent γ is smaller than 3 [60]. Furthermore, in some situations, $\overline{\ell}(N)$ even approaches a constant as $N \to \infty$. This is the case, for example, if a single node in a network attracts a finite number of all connections, as in Fig. 2.7. Unfortunately, it is hardly possible to distinguish between various slowly varying dependences on N in finite real-world networks.

[5] Recall that $\overline{b} = (\langle q^2 \rangle - \langle q \rangle)/\langle q \rangle$.

Not everything can be done in the framework of the convenient tree ansatz. Here we have indicated only one problem which cannot be solved within this approximation. Let us introduce an important notion which helps to characterize the distribution of shortest paths over a network. *The betweenness centrality* (physicists also call it *load*) shows how often shortest paths in a network pass through a given node.[6] The betweenness centrality of a given node is proportional to the relative number of shortest paths between other nodes which run through this node.[7] More rigorously, the betweenness centrality of a node is defined as follows. Let $s(i,j) > 0$ be the number of shortest paths between nodes i and j, while $s(i,v,j)$ is the number of these paths passing through node v. Then the betweenness centrality $B(v)$ of node v is

[6] The notion of betweenness centrality was first proposed in sociology.

[7] Similarly, one can introduce the betweenness centrality of a link.

$$B(v) \equiv \sum_{i,j \neq v} \frac{s(i,v,j)}{s(i,j)} \,, \tag{5.4}$$

where the sum is over all nodes other than node v. Similarly to the degree distribution, for a random network, one should introduce the betweenness centrality distribution $\mathcal{P}(B)$.

Figure 5.3 shows schematically all possible configurations of shortest paths between three nodes in the same connected component. Note configuration (c) which implies that the presence of loops in a network certainly changes the value of betweenness centrality. Therefore, to calculate the betweenness centrality distribution, one has take into account loops. It turns out that this is a difficult task, and so the problem for loopy networks is still open. As far as we know (2008), even for classical random graphs, an exact betweenness centrality distribution has not been found. Nonetheless, this distribution has been investigated in numerous empirical works and numerical simulations. In a wide range

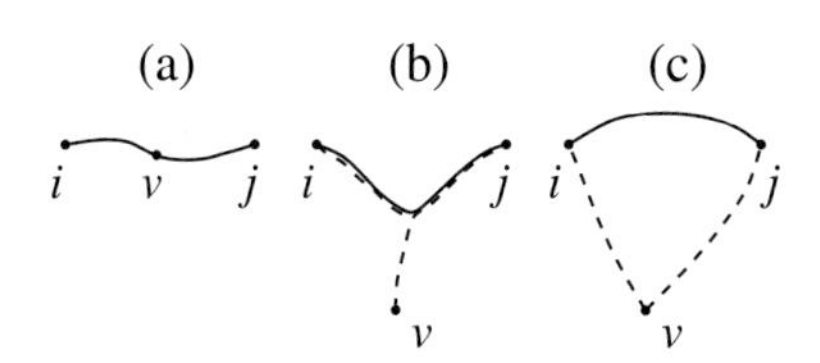

Fig. 5.3 Possible configurations of the shortest paths connecting nodes i, j, and v. The dotted lines show the shortest paths between nodes i and v, j and v.

of networks, including scale-free nets, the observed distribution is rather close to $\mathcal{P}(B) \sim B^{-2}$ [97].

In uncorrelated networks, the width of the distribution of internode distances was found to be independent of N. If a network is sufficiently small, then a noticeable fraction of nodes may be separated by distances essentially greater than $\overline{\ell}$. The statistics for this remote part of a network is of particular interest. The relevant quantity is the distribution of the number of the ℓ-th nearest neighbours of a randomly chosen node if $\ell \gg \overline{\ell}$. It is even more convenient to use the very close distribution of the number of nodes $z_{>\ell}$ at a distance greater than ℓ from a node. Interestingly, a form of these distributions is essentially determined by the presence (or absence) of dead ends in a network. If there are no nodes with a single link in a network, then these distributions rapidly decay. If a finite fraction of all nodes are dead ends, then these distributions decay quite slowly, as $z_{>\ell}^{-2}$.

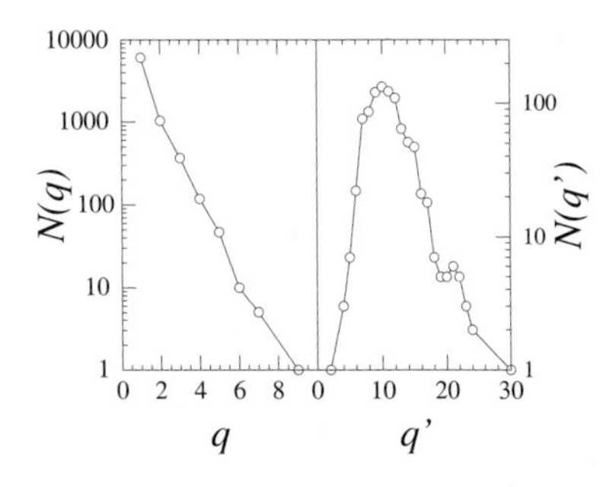

Fig. 5.4 Empirical statistics of the bipartite Fortune 1000 graph. Frequency with which a director sits on q board is on the left panel; frequency with which q' directors sit on a board is on the right panel. Adapted from Newman, Strogatz and Watts [140].

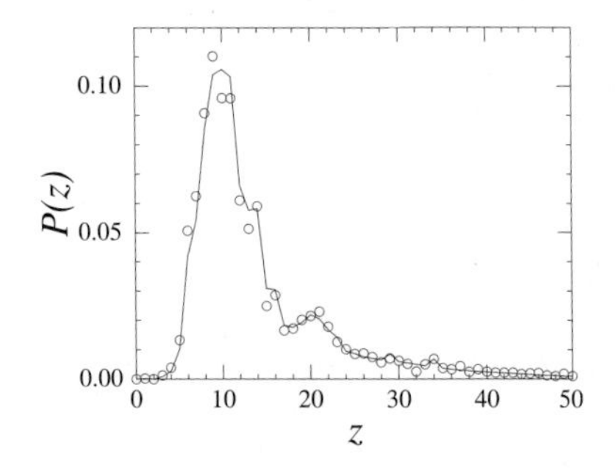

Fig. 5.5 Statistics of the one mode projection (to directors) of the bipartite Fortune 1000 graph. Dots are the empirical data: probability that a director has in total z co-directors on all his boards. The solid line was calculated for the one-mode projection of an uncorrelated bipartite graph with the same statistics as in Fig. 5.4. Adapted from [140].

5.6 Uncorrelated bipartite networks

We have demonstrated how to build an uncorrelated undirected one-partite network. Similarly, one can define uncorrelated models of other networks: directed networks, multi-partite networks, etc. A number of examples were described in a seminal paper by Newman, Strogatz and Watts (2001) [140]. For demonstration purposes, let us focus on undirected bipartite networks, see Fig. 1.4 (a). If the reader needs a real-world example, it can be a bipartite collaboration network of movie actors or one of the networks of the members of boards of directors. In the movie actor network, one type of node is actors and the other is movies where they played. In the network of directors, one kind of node is directors and the other is boards on which they sit. An uncorrelated bipartite network is a network with (i) given numbers of nodes of each sort, N_1 and N_2 and (ii) given degree distributions $P_1(q)$ and $P_2(q')$ for each kind of node, otherwise they are uniformly random. As usual, we can also say that this is the maximally random bipartite network with given N_1, N_2, $P_1(q)$, and $P_2(q')$. Similarly to the one-partite uncorrelated networks, this network has few short loops. Remarkably, the one-mode projection of an uncorrelated bipartite network, Fig. 1.4 (b), is already correlated. Furthermore, this projection is already not tree-like—it has many short loops. In particular, the clustering coefficient C of the one-mode projection does not vanish as the size of this network approaches infinity. This shows how one can easily get large clustering starting from an uncorrelated network.

How far is this model from real bipartite networks? It turns out that it sometimes describes a real situation surprisingly well. In their work, Newman, Strogatz and Watts inspected the Fortune 1000 graph, namely a bipartite network of the members of boards of directors of the US companies with the highest revenues. They obtained three empirical degree distributions: two degree distributions for each of two kinds of nodes, Fig. 5.4, and the degree distribution to the one-mode projection of this graph to the set of nodes-directors, the dots in Fig. 5.5. Then they built an uncorrelated bipartite network with the same distributions $P_1(q)$ and $P_2(q')$ as in the real network. It was easy to compute the degree distribution of the one-mode projection of this model network, the solid line in Fig. 5.5. The reader can see that the result is very close to the real-world data. The calculated clustering coefficient of the one-mode projection, $C = 0.59$, practically coincided with the measured one. Note the large value of the clustering coefficient. One has to admit that such close agreement is the exception rather than the rule.

Percolation and epidemics

6

The uncorrelated models provide us with a discovery tool to explore complex networks. With this tool, in this lecture we approach a set of related problems for complex networks. What is the global organization of these networks? How can the networks be destroyed? How do epidemics spread through complex networks?

6.1 Connected components in uncorrelated networks

We have described the system of connected components in classical random graphs, see Fig. 2.4. In essence, this qualitative picture with a single giant and numerous finite connected components is quite general. The problem is that in different complex networks, the giant component can emerge in very different ways, with differently distributed finite components. Between 1995 and 1998, graph theory mathematicians Michael Molloy and Bruce A. Reed found a way to explicitly obtain the statistics of connected components in the configuration model [128, 129]. Here we explain the idea of this basic approach which physicists mostly know from the previously mentioned paper of Newman, Strogatz, and Watts (2001). Many key results in the science of complex networks have been obtained using this approach.

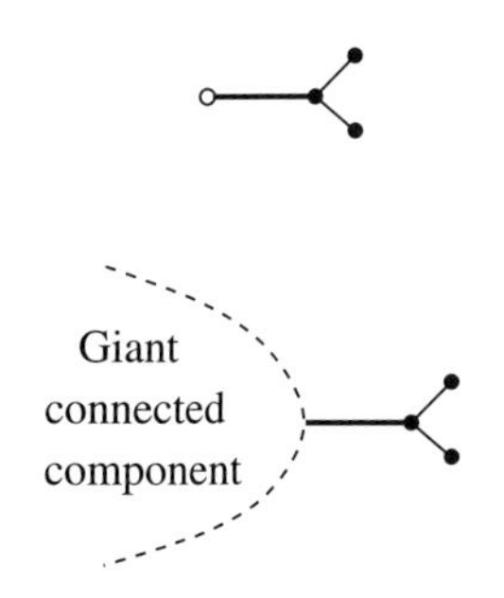

Fig. 6.1 Explanation of the probability x. The right end of the horizontal link is selected. Both of these two, at first sight very different configurations, contribute to x.

Let us, for example, find a giant connected component in the configuration model of an infinite undirected network with a degree distribution $P(q)$. Recall that in this model, (i) all nodes and links are statistically independent, (ii) the degrees of the nearest neighbours are independent of each other, (iii) the degree distribution of an end node of a link, coinciding with the degree distribution of any nearest neighbour of a node, is $qP(q)/\langle q\rangle$, and (iv) the structure of a network is locally tree-like.

These properties allow us to introduce the following basic probability x in this problem. Choose at random a link. Select with equal probability one of its ends. Then x is the probability that strictly following this link in the direction of the chosen end we arrive at a 'finite connected component'. In this 'finite component', the nodes approachable following the link in the opposite direction are ignored. So this 'finite component', in principle, may belong to the giant connected component of the network. Figure 6.1 explains this definition. One can graphically represent probability x as shown in Fig. 6.2. This is simply a useful notation.

$x =$ —○

Fig. 6.2 Graphical notation for probability x.

Using this notation, we can conveniently represent various characteristics of a network. For example, the probability that a randomly chosen link belongs to one of the finite connected components has a graphical representation shown in Fig. 6.3. Indeed, in this situation, following the link in any direction we must approach a finite component. Clearly, this probability equals x^2. Then the fraction of links belonging to the giant connected component equals $1 - x^2$. This provides us with a way to get x if we succeed in measuring the number of links in the giant component of an uncorrelated network. Our aim, however, is not to measure but to calculate characteristics of connected components for a given $P(q)$.

$= x^2$

Fig. 6.3 The probability that a link is in one of the finite connected components.

If we find a way to get x, we will be able to obtain the relative size S of the giant connected component and many other quantities. Figure 6.4 shows how to find the probability $1 - S$ that a randomly chosen node belongs to one of the finite components, if x is known. This is a sum of the probabilities: the probability that a node has no connections plus the probability that it has a single link and that this link leads to a finite component, and so on. Note that this equality necessarily implies a tree-like structure of a network. Figure 6.4 leads to the following simple formula:

$$1 - S = \sum_q P(q)x^q \,. \tag{6.1}$$

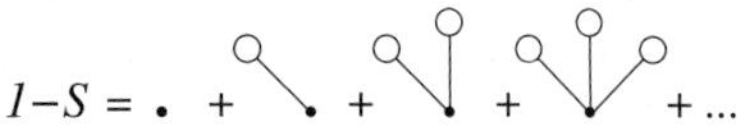

$1-S = \cdot + \quad + \quad + \quad + \ldots$

Fig. 6.4 The probability $1 - S$ that a randomly selected node belongs to one of the finite connected components. Here S is the relative number of nodes in the giant connected component.

Now we should find probability x. To derive an equation for x we again use graphical notations. Figure 6.5 explains the form of this equation. To write it down we use the degree distribution of an end node of a randomly chosen link, $qP(q)/\langle q\rangle$. As a result we have

$$x = \sum_q \frac{qP(q)}{\langle q\rangle} x^{q-1} \,. \tag{6.2}$$

Thus the problem is essentially reduced to the analysis of this equation for a given distribution $P(q)$. When it has a non-trivial solution $x < 1$, the network has a giant connected component. One can easily show that this takes place when the mean number of second nearest neighbours of a node exceeds the mean number of nearest neighbours: $z_2 > z_1$. This is the celebrated *Molloy–Reed criterion* (1995),

$$\langle q^2\rangle - \langle q\rangle > \langle q\rangle \tag{6.3}$$

which is the mean branching $\overline{b} > 1$. At the birth point of a giant connected component, the mean number of second nearest neighbours coincides with the mean number of nearest neigbours, $z_2 = z_1$. In particular, this leads to the proper critical point $\langle q\rangle = 1$ for classical random graphs.

$= \quad + \quad + \quad + \quad + \ldots$

Fig. 6.5 Graphical representation of the equation for probability x.

Surprisingly, these equations and the criterion may still be valid even when the second moment of a degree distribution diverges and the tree ansatz becomes dubious. Look at criterion (6.3): if $\langle q^2\rangle$ diverges, a network necessarily has a giant connected component. On the other hand, one can check using these equations that, if an uncorrelated network has no dead ends, then almost all nodes in the network are in the giant connected component. In the next section we will show how strongly the size of the giant component depends on the form of the degree distribution.

Using these equations, one can find the organization of connections inside the giant component. Importantly, the distribution of connections in a giant connected component may essentially differ from that of the entire network. The strongest difference is observed close to the birth point, where the degree distribution in a giant component is proportional to $qP(q)$.

The same ideas were successfully used to find the statistics of finite connected components in uncorrelated networks [140]. The derivation, however, was more cumbersome, so here we only outline the main results. Similarly to classical random graphs, in any uncorrelated network the size distribution of finite components, $\mathcal{P}(s)$, is a power law at the birth point of a giant connected component. If the degree distribution of a network decays sufficiently rapidly, then the distribution $\mathcal{P}(s) \sim s^{-5/2}$ at this critical point, as in classical random graphs. For slowly decaying degree distributions, the exponent of $\mathcal{P}(s)$ differs from 5/2.[1]

That was a 'critical' distribution. What about the size distribution away from the critical point? When a network with any degree distribution contains a giant connected component, its finite components have exponentially rapidly decaying size distribution.[2] For 'non-critical' situations, this is not surprising but typical. What is surprising is that in the other phase, where a giant connected component is absent, scale-free and non-scale-free networks have very different distributions of finite components. In scale-free uncorrelated networks in this situation the size distribution decays slowly, $\mathcal{P}(s) \sim s^{-\gamma}$ [136]. In contrast, in non-scale-free networks ($\gamma = \infty$), $\mathcal{P}(s)$ decays exponentially rapidly.

[1] In particular, if an uncorrelated network is scale-free, $P(q) \sim q^{-\gamma}$, then (i) for $\gamma > 4$, the distribution $\mathcal{P}(s) \sim s^{-5/2}$, and (ii) as γ decreases from 4 to 3, the exponent of $\mathcal{P}(s)$ increases from 5/2 to 3 [61].

[2] This is true for an arbitrary network, and not only for uncorrelated.

6.2 Ultra-resilience phenomenon

One of main questions to ask about a network concerns its robustness. The standard approach to this problem is to study how the removal of nodes influences the global structure and function of a network. Remove a fraction of nodes from a network and investigate how the size of a giant connected component diminishes—this is the essence of hundreds of papers on this topic. In terms of condensed matter theory, this is a site percolation problem.

The nodes for deletion may be chosen in various ways. The simplest way is to choose them uniformly at random. Let us remove a fraction $1 - p$ of uniformly selected nodes from an uncorrelated network with a given degree distribution, and so with given $z_1 = \langle q \rangle$, $z_2 = \langle q^2 \rangle - \langle q \rangle$, and $\overline{b} = z_2/z_1$. Then, when will a giant connected component disappear? The answer was found using the same approach as in the preceding section. The giant connected component is present in a network if

$$pz_2 > z_1 , \tag{6.4}$$

so the percolation threshold (the birth point of a giant connected component) is at

$$p_c = \frac{z_2}{z_1} = \frac{\langle q \rangle}{\langle q^2 \rangle - \langle q \rangle} = \frac{1}{\overline{b}}\,. \tag{6.5}$$

Here z_1, z_2, $\overline{b}$, $\langle q \rangle$, and $\langle q^2 \rangle$ are for an undamaged network.[3] This relation shows that if the degree distribution of an uncorrelated network has so heavy a tail that $\langle q^2 \rangle$ diverges, then the threshold p_c is zero. In other words, we have to remove practically all nodes from this network to eliminate its giant connected component. In scale-free networks, $\langle q^2 \rangle$ diverges when exponent $\gamma \leq 3$, which is the case for many real-world networks. In 2000, Albert, Jeong, and Barabási investigated a number of real networks, including the WWW and the Internet, and for all of them observed this ultra-resilience against random failures [6].

[3] For derivations and details see papers [50, 58].

Often the ultra-resilience is considered as one of the most impressive phenomena induced by complex network architectures. However, this effect of the heavy tails of degree distributions (the abundance of hubs) is almost trivial. Let us discuss the extreme situation—a finite fraction c of nodes are connected to a single node. For simplicity, we assume that without this super-hub, all of these cN nodes are not connected. Suppose that we select nodes for removal uniformly randomly, with probability $1-p$. Then the average size of a giant connected component in this network is $(1-p)\cdot 0 + p\cdot pcN = p^2cN$. So for any non-zero p, the giant connected component contains a finite fraction p^2c of nodes in the network, which means ultra-resilience. Now suppose that we want to cause the maximum damage. Then, as is natural, remove the hub and there will be a chance to destroy a giant component. Thus networks with heavy degree distributions are extremely robust against uniformly random damage but, simultaneously, quite weak against intentional damage. It is usually enough to remove a few per cent of highly connected nodes to completely split a network with a heavy-tailed degree distribution [50, 59]. This is in sharp contrast to the networks with rapidly decreasing distributions of connections, where both kinds of damage produce similar effects. Irrespective of its simplicity, the ultra-resilience phenomenon have numerous consequences with regard to the functioning of complex networks, which we will discuss in the following lectures.

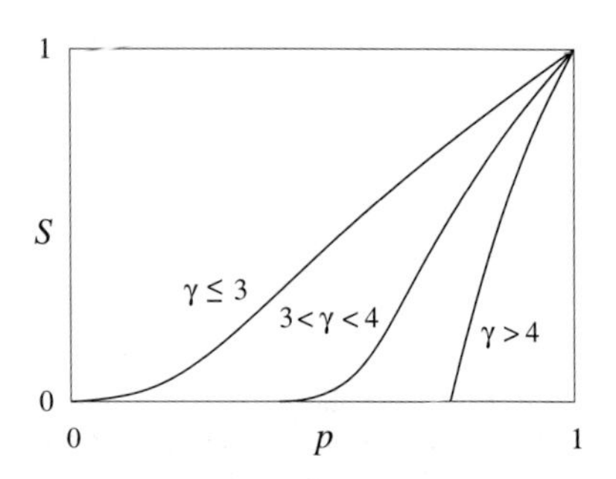

Fig. 6.6 Relative size of a giant connected component versus the fraction of nodes retained after random damage. The three curves show the typical dependences at various values of the exponent of a scale-free degree distribution.

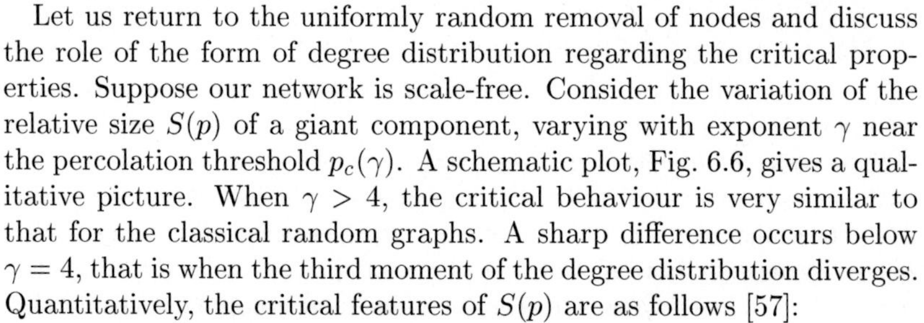

Let us return to the uniformly random removal of nodes and discuss the role of the form of degree distribution regarding the critical properties. Suppose our network is scale-free. Consider the variation of the relative size $S(p)$ of a giant component, varying with exponent γ near the percolation threshold $p_c(\gamma)$. A schematic plot, Fig. 6.6, gives a qualitative picture. When $\gamma > 4$, the critical behaviour is very similar to that for the classical random graphs. A sharp difference occurs below $\gamma = 4$, that is when the third moment of the degree distribution diverges. Quantitatively, the critical features of $S(p)$ are as follows [57]:

(i) if $\gamma > 4$, then $S \propto p - p_c$,
(ii) if $3 < \gamma < 4$, then $S \propto (p - p_c)^{1/(\gamma-3)}$,
(iii) if $\gamma < 3$, then $S \propto p^{1+1/(3-\gamma)}$ and $p_c = 0$.

These results were obtained by using the approach described in the preceding section. They assume that any finite neighbourhood of a node has no loops. Items (ii) and (iii) in this list must surprise physicists. Let us discuss this point in more detail.

As is well known from condensed matter physics, critical phenomena in models defined on the top of infinite-dimensional objects are exactly described by mean-field (molecular-field) theories.[4] There are three standard infinite-dimensional objects in condensed matter: (i) an infinite-dimensional lattice, (ii) a fully connected graph, and (iii) a Bethe lattice. For percolation on each of these objects, $S(p) \propto p - p_c$, and this is also the result of mean-field theory for sufficiently homogeneous systems. This theory, however, is not applicable to low-dimensional systems. It fails, for example, in two- and three-dimensional lattices. One of the challenges of theoretical physics in the 20th century was to advance beyond mean-field theories. During several of the last decades of the 20th century physicists arrived at a clear understanding of critical phenomena in low-dimensional ordered and disordered lattices in situations where the mean-field theories fail. In these situations, critical features of observables are power-law singularities, e.g., $S(p) \propto (p-p_c)^\beta$, with non-mean-field exponents. The birth of a giant component in classical random graphs and networks with rapidly decaying degree distributions is well described by standard mean-field theory, as it should be for small worlds. Our list, however, shows that when $\gamma < 4$, the standard mean-field theory fails. For γ between 3 and 4, the critical feature of $S(p)$ differs from the standard mean-field one as in two- and three-dimensional lattices, though our networks are infinite-dimensional. On the other hand, it turns out that for any value of exponent γ, the mean size of a finite component to which a node belongs has the same critical singularity: $\langle s\rangle' \propto 1/|p - p_c|$. So for this quantity, in contrast to $S(p)$, the standard mean-field works at any degree distribution.

[4] In physics, a mean-field theory is the main approach to systems of interacting agents, for example, spin $\mathbf{S}$. In this approximate approach, instead of solving a full set of equations for all spins, a single spin in a mean field from its neighbours is considered. This effective field is taken as $\mathbf{H}_{\text{eff}} \approx \langle\mathbf{S}\rangle z_1$, i.e. the mean value of spin times the coordination number of a lattice. Within this approximation, one can obtain $\langle\mathbf{S}\rangle = f(\mathbf{H}_{\text{eff}}) = f(\langle\mathbf{S}\rangle z_1)$ and then find $\langle\mathbf{S}\rangle$ from this self-consistent equation.

At first sight, this combination of singularities of different kinds looks very strange. The actual reason for this 'strange' critical behaviour is the strong heterogeneity of these complex networks, namely a wide range of node degrees. This heterogeneity does not allow one to use the standard mean-field theory developed for homogeneous systems. Nonetheless, if we modify a mean-field theory, accounting for the heterogeneity of a network, then we will get correct critical singularities.[5]

[5] We can even treat eqn (6.2) as a self-consistent equation for order parameter in a specific mean-field theory.

6.3 Finite-size effects

The reader has certainly noticed that the strongest effect takes place when the second moment of the degree distribution diverges. The problem is that this divergence is possible only if a network is infinite, while all real networks are, of course, finite. In finite networks any heavy-tailed degree distribution necessarily has a cut-off—an end part of the distribution with a rapid decay, see Fig. 6.7. The position of this cut-off, $q_{\text{cut}}(N)$, determines the values of higher moments of a degree

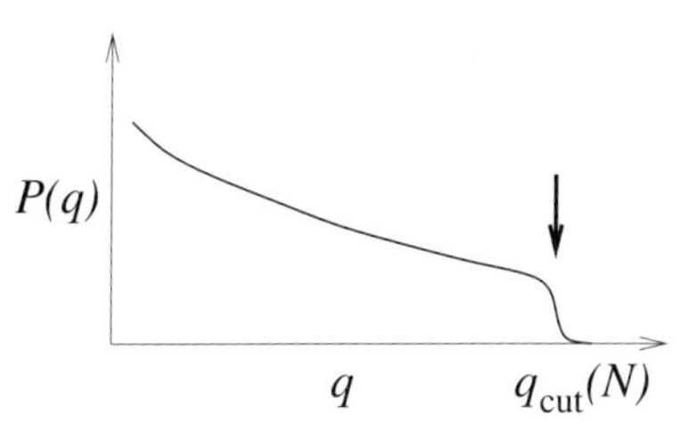

Fig. 6.7 Size-dependent cut-off of a degree distribution.

distribution and so strongly influences various properties of a network. This is why $q_{\rm cut}(N)$ is one of the basic network characteristics. For example, in a scale-free network with exponent γ less than 3, the second moment of a degree distribution takes the form $\langle q^2 \rangle \sim q_{\rm cut}^{3-\gamma}(N)$. Size dependent $q_{\rm cut}(N)$ differs from model to model.[6]

[6] See the list of cut-offs for various network models in review [77].

For a single scale-free graph, that is for a single member of an ensemble, the highest possible cut-off $q_{\rm cut}(N)$ is estimated to be of the order of $N^{1/(\gamma-1)}$. If the cut-off is of this order, then the graph has a finite number of nodes with degrees exceeding $q_{\rm cut}$, or even a single node in this range. This cut-off is called 'natural'. In some network models, however, $q_{\rm cut}$ is far below that 'natural' value. Here we indicate only one network model with a cut-off of this kind. Consider a scale-free uncorrelated network with exponent $\gamma < 3$, where multiple connections and one-loops are forbidden. One can show that with these restrictions, the cut-off $q_{\rm cut}(N)$ cannot be higher than of the order of $N^{1/2}$. Nodes of higher degrees would make this network correlated. Remarkably, if nodes with this (of the order of $N^{1/2}$) or higher degrees are present, then they form a densely interconnected club within a sparse network. Thus nodes with the highest numbers of links are interconnected. This was named a *rich-club phenomenon* and observed in many real-world networks, including the AS and router level Internet networks [185]. Note, however, that this club contains only a tiny fraction of nodes in a network.[7]

[7] The reader can easily check that this elite club contains only about $N^{(3-\gamma)/2}$ nodes [63]; which is much less than N.

To estimate the size effect, one can formally substitute the resulting second moment of the degree distribution in a finite network into eqn (6.5) for the percolation threshold and readily obtain a finite p_c. Any physicist will understand that this is a very naive approach. Indeed, continuous phase transitions are, in principle, impossible in finite systems, and $p_c(N)$ and a giant connected component are well defined only in the limit $N \to \infty$. Nonetheless, this dubious way is possible if we only need a simple estimate. Thus, in real-world networks, the heavy tails of degree distributions cannot cause absolute resilience. We can claim, however, that even finite networks of this architecture are far more robust against random failures than the classical random graphs.

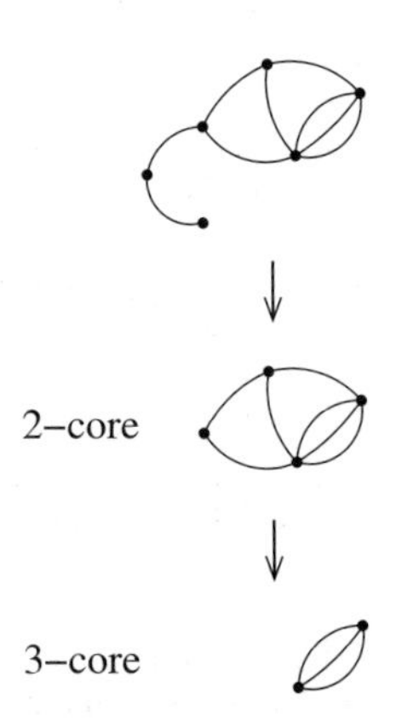

Fig. 6.8 The k-cores of a small graph. To find a k-core, one can use the 'pruning algorithm'. Remove from a graph all nodes of degree less than k. After that, some of the remaining nodes may occur with less than k links. Prune these nodes, and so on. The final result, if it exists, is the k-core.

6.4 k-cores

The k-core of a network is its largest subgraph whose nodes have at least k connections (within this subgraph, of course) [54]. Figure 6.8 shows the k-cores in a small graph. In general, there is a set of successively enclosed k-cores, similarly to a Russian nesting doll—'matrioshka'. The full set of the k-cores of a network, $k = 1, 2, 3, \ldots$, presents essentially a more detailed description of the network than the one based on connected components. Knowledge of k-cores enables us to indicate the best interconnected parts in a network. The k-cores have been used as a network visualization tool. The reader will find the image gallery of picturesque k-core decompositions of various networks in [9] and on a special site http://xavier.informatics.indiana.edu/lanet-vi/.

In essence, the 2-cores are rather close to connected components—only the dangling chains are eliminated, see Fig. 6.8. The 3- and higher-k-cores, however, show a sharp contrast to connected components. Let us discuss these k-cores.

First note that the trees have no ($k{\geq}2$)-cores. Use pruning, similarly to Fig. 6.8, to understand this claim. Then if a network is locally tree-like, it, in principle, cannot have finite ($k{\geq}2$)-cores. In loopy networks (more precisely, networks having short loops), a single giant and numerous finite k-cores can coexist, but in locally tree-like networks there can only be a single giant k-core. In 1979, Chalupa, Leath, and Reich found that the birth of this giant k-core, for $k \geq 3$, is a quite unusual phase transition [54], different both from continuous and first-order transitions, see Fig. 6.9. Formally speaking, these authors focused not on networks, but rather on lattices. Namely, they studied the process of the elimination of a giant k-core by removing random sites from a lattice. Their findings, however, turned out to be valid for a wide range of lattice and network architectures.[8]

[8] In their calculations, these researchers used a Bethe lattice, which we know is infinite-dimensional.

Normally, physicists divide phase transitions into continuous and first order, Fig. 6.9 (a) and (b). The birth of a giant connected component is a continuous transition. There is no jump of S (relative size of a giant component) at the birth point, Fig. 6.9 (a). In contrast, in a first-order phase transition, an order parameter X emerges abruptly, Fig. 6.9 (b). Surprisingly, the transition associated with the birth of a k-core combines the features of continuous and first-order phase transitions, see Fig. 6.9 (b). (i) At the birth point, q_c, there is a jump of S_k, the relative size of a k-core, from zero to a finite value. (ii) In addition, there is a square root singularity at the same point:

$$S_k(\langle q\rangle) - S_k(q_c) \cong \text{const}\sqrt{\langle q\rangle - q_c}\,. \tag{6.6}$$

This phase transition, which is called *a hybrid transition* or, sometimes, a mixed transition, is not so common as first- and second- order phase transitions. Note, however, that k-cores have many applications, for example, in various jamming phenomena and in systems for limiting metastable states.

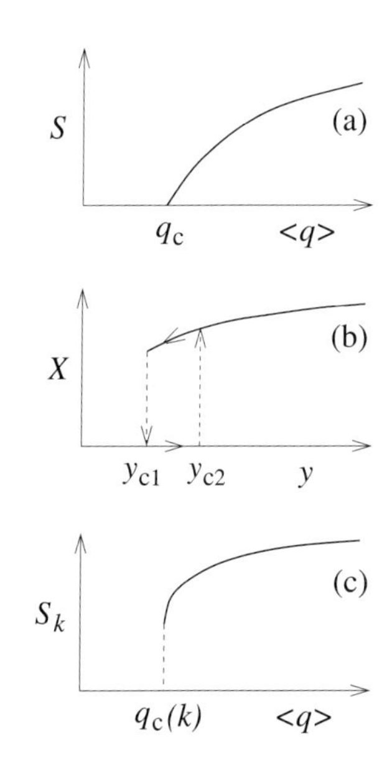

Fig. 6.9 Three phase transitions. (a) Continuous transition, the birth of a giant connected component. S is the relative size of a giant component. (b) First-order transition. X stays for an order parameter, y denotes a control parameter. (c) Hybrid phase transition, the birth of a k-core. Here S_k is the relative size of a giant k-core. Near the jump at $q_c(k)$, the deviation of S_k varies as $\sqrt{\langle q\rangle - q_c(k)}$.

Here we will not discuss the nature of the hybrid phase transition, see discussion in review [77]. For us, it is more interesting how k-cores are organized in various networks. It was found that this specific transition takes place in networks with sufficiently rapidly decaying degree distributions, namely if $\langle q^2\rangle < \infty$, see schematic Fig. 6.10 (a). If the degree distribution of an infinite network decays more slowly (in scale-free networks, this corresponds to the region $\gamma \leq 3$), then this network has an infinite sequence of k-cores for any non-zero value of mean degree, Fig. 6.10 (b). In this range of γ, the k-cores behave similarly to a giant connected component. Instead of mean degree, we can use another control parameter, p, which is the probability that a node is retained in a uniformly damaged network. On a qualitative level, we simply substitute $\langle q\rangle$ in Fig. 6.10 for p. Then compare Figs. 6.6 and 6.10. Clearly, for $\gamma \leq 3$ in an infinite network, all k-cores are ultra-resilient against

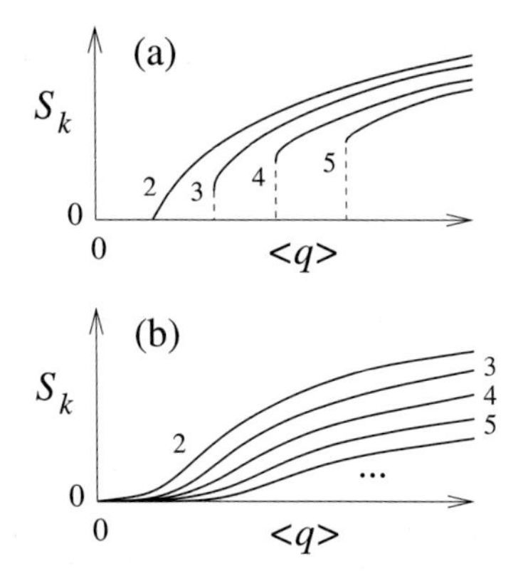

Fig. 6.10 Schematic view of the birth process of the k-core set in uncorrelated networks when: (a) the second moment of a degree distribution is finite, (b) the γ exponent of a power-law degree distribution does not exceed 3 [76]. Instead of mean degree $\langle q \rangle$, one can use a concentration of retained nodes in a uniformly damaged network.

random damage. If a network is finite, the k-cores are already not absolutely robust—Fig. 6.10 (b) transforms into Fig. 6.10 (a). Notice that as random damage increases (p diminishes), first the highest k-core disappears, then the second highest, and so on. That is, the damage primarily spoils the best interlinked areas in a network.

6.5 Epidemics in networks

How does an infectious disease spread? Generations of epidemiologists and population biologists have studied this challenging problem. One can safely say that the spread of disease in homogeneous systems is now well understood. Traditional theories describe epidemics within fully connected graphs, classical random graphs, and lattices. Usually in epidemic models, each of the individuals can be in two or three states: S—susceptible, I—infective, and R—removed or recovered but not susceptible. So the dynamics of epidemics should be described in terms of three probabilities for each of the individuals in a population: the probabilities $i(t)$, $s(t)$, and $r(t)$ that this individual is infective, susceptible, and removed, respectively. Specific epidemic models use different sets of states and set rules for transitions between them.

To be more concrete, in *the SIS model*, an individual can be susceptible or infective. The rate of recovery of an infected individual ($I \to S$) is μ. A susceptible individual becomes infected ($S \to I$) with a rate β if at least one of its nearest neighbours is infective. The main parameter of the problem is the so-called reproductive number $\lambda = \beta/\mu$.

In the second basic model—*the SIR model*—an individual can be susceptible, infective, or recovered. The 'recovery' rate for an infected individual ($I \to R$) is μ.[9] As in the SIS model, a susceptible individual becomes infected ($S \to I$) with a rate β if at least one of its nearest neighbours is infective. The main parameter of the problem is the reproductive number for this infection, $\lambda = \beta/\mu$.

[9] This 'recovery' may occur as removal.

These models are defined on various substrates—networks or lattices—where each individual permanently occupies its own node and never changes it. Let us first touch upon classical homogeneous situations, where the nodes in a network or a lattice have a narrow distribution of connections.[10] Suppose that initially a single or a few nodes are infected. It turns out that the disease will quickly die out if the reproductive number is below some value, *an epidemic threshold*, λ_c. This is one of the fundamental notions in epidemiology. Another one is *the prevalence*, which is the fraction of infected individuals. In homogeneous situations, the epidemic threshold is determined by the mean degree of a node in a network: $\lambda_c \sim 1/\langle q \rangle$. Above the epidemic threshold, an epidemic spreads throughout a network.

[10] Here we discuss only the spread of disease within a giant connected component.

This general picture is valid for the SIS and SIR, and for many other epidemic models. There are differences, but they are not that important. Above the epidemic threshold in the SIS model, the prevalence $\langle i \rangle(t)$ in a population grows monotonously. It is finite in the final state. In contrast to this, in the SIR model, $\langle i \rangle(t)$ shows an epidemic outbreak but approaches zero at $t \to \infty$: $\langle i \rangle = 0$. This outbreak, however, results in a finite fraction of removed (or recovered) individuals in the final state, $\langle r \rangle > 0$. Figure 6.11 shows the evolution of an epidemic in these two models.

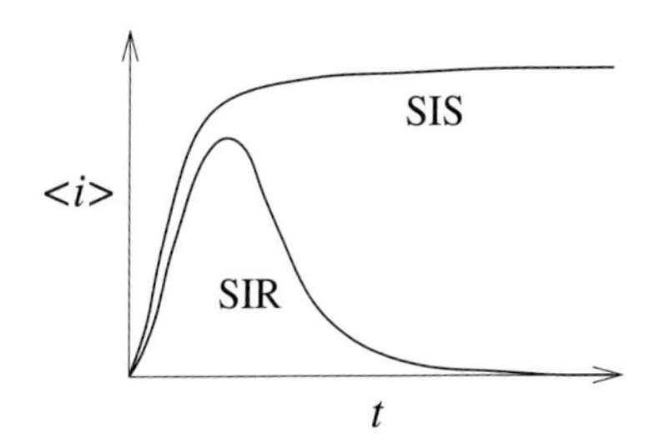

Fig. 6.11 Fraction of infected individuals versus time above an epidemic threshold in the SIS and SIR models. In the SIR model, the fraction of removed individuals $\langle r \rangle$ grows monotonously with time and approaches a constant value at $t \to \infty$.

In 2001, Pastor-Satorras and Vespignani [149] moved beyond this traditional focus on epidemics in homogeneous media. They considered the speed of infectious disease within an uncorrelated network with an arbitrary degree distribution. The results that they obtained naturally generalize the picture of epidemics described above. The most important result was the epidemic threshold. In the SIR model,

$$\lambda_c = \frac{\langle q \rangle}{\langle q^2 \rangle - \langle q \rangle} = \frac{1}{\bar{b}}, \qquad (6.7)$$

which coincides with the percolation threshold, eqn (6.5). For physicists, this coincidence is not surprising. They know that the SIR model is equivalent to the percolation problem in many respects.[11] Consequently, in networks with heavy-tailed degree distribution, the epidemic threshold is low. In real-world networks, it may be dramatically lower than the $1/\langle q \rangle$ value of classical random graphs. What about the dynamics of epidemics above the epidemic threshold? In the qualitative Fig. 6.11 for the evolution of prevalence, the characteristic time scale is $\tau = \lambda_c/\lambda$. This is a typical time for an epidemic outbreak.

[11] For the SIS model, a close expression was found, $\lambda_c = \langle q \rangle / \langle q^2 \rangle$.

Thus the presence of hubs in networks leads to low epidemic threshold and τ. How can we protect these networks against epidemics, that is, increase the epidemic threshold? The obvious answer is primarily to immunize the strongly connected nodes. Unfortunately, for this it is necessary to know these nodes, which is often not the case. This is why a number of immunization strategies were proposed on how to pick up the 'best' individuals for immunization with restricted information about a network [62]. The weak point of these algorithms is that usually they assume some type of the network architecture. If, for example, we suppose that a network is uncorrelated, than a relevant strategy appears as follows. Choose a node at random and immunize one of its nearest neighbours. Repeat this as many times as you like. If the real network is indeed uncorrelated, then this strategy certainly works—you will immunize the hubs with higher probability. However, the same strategy, applied to networks of another type, may be a complete failure.[12]

[12] If you want to choose a set of nodes with degree distribution $qP(q)/\langle q \rangle$ from a network with an unknown architecture, use the following approach. Choose a node uniformly from the set of all nodes. Then select each of its nearest neighbours with some probability p, where $0 < p \leq 1$. Repeat this for as many randomly chosen nodes as you want.

In many respects, the basic infection models, which we have discussed, are not realistic. In these models individuals stay permanently at their nodes, one individual per node. In the real world, the spread of infectious disease, like pandemic influenza, is rather due to high population mobility. Millions of passengers travel across transportation networks, from city to city, increasing the potential for rapid spread of disease. In models for this process, nodes—urban areas—are occupied by many individuals. With some mobility rate κ, these individuals travel from node to node with a chance to transmit infectious disease from one population to another. (Within populations, the spread of disease is described by standard epidemic models.) In other words, these metapopulation models describe the process of the invasion of disease in a network of well-interlinked populations. Remarkably, the main conclusions for epidemics obtained using the SIS and SIR models remain valid in metapopulation models. If these models are defined on top of uncorrelated networks, an invasion threshold for an epidemic, κ_c, is proportional to $\langle q \rangle / (\langle q^2 \rangle - \langle q \rangle) = 1/\overline{b}$, similarly to the percolation and epidemic thresholds (6.5) and (6.7) [64].

In standard epidemic models, a final stationary prevalence $\langle i \rangle$ emerges at the epidemic threshold in a similar way as a giant connected component at the percolation threshold, see Fig. 6.6. Similarly to a giant connected component, $\langle i \rangle$ rises continuously from zero as is should be at a continuous phase transition. There is a class of models, however, where prevalence emerges explosively, after a sudden jump from zero. In the best known of these models, *the bootstrap percolation problem*, disease spreads in the following way [3].

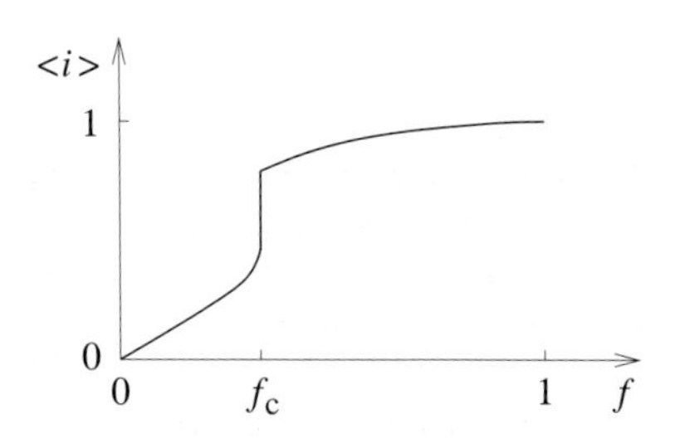

Fig. 6.12 Schematic plot of a final prevalence as a function of the fraction f of initially infected nodes in the bootstrap percolation problem. Note a square-root singularity on the low-f side of the transition: $\langle i \rangle (f_c - 0) - \langle i \rangle (f) \propto \sqrt{f_c - f}$.

- At the initial moment, a fraction f of nodes in a network are infected for ever. Then, at each time step, all nodes with k or more infected nearest neighbours become infected. Here $k \geq 3$.

Figure 6.12 demonstrates how a final prevalence, resulting from this process, depends on f. This picture is valid for classical random graphs. There is a discontinuous jump at the critical point f_c accompanied by a square-root singularity on the left-hand side of the transition. So this transition is hybrid similarly to the k-core transition. This is not strange, because the bootstrap percolation and k-core problems are generically related. Recall the definition of a k-core and compare somewhat complementary Figs. 6.12 and 6.9(c). Notice that the square-root singularities in these plots are on opposite sides of the transitions. Finally, it is worthwhile mentioning that the main applications of bootstrap percolation are not to epidemiology, but rather to the failure of units in large arrays and related problems.

Self-organization of networks

7

The formal network constructions (the configuration model and others) which we have discussed do not explain why and how networks get their complex structures. In this lecture we show how the evolution of a network shapes its architecture.

7.1 Random recursive trees

Let us start with the simplest growing random networks. We have already introduced random recursive trees in Lecture 3. Recall the growth process illustrated by Fig. 3.6 (b): at each time step, a new node is attached to a uniformly randomly chosen existing node by a single link—uniform random attachment. We explained that these networks differ strongly from equilibrium ones. In particular, their degree distribution, $P(q) = 2^{-q}$ decays more slowly than the Poisson degree distributions of the equilibrium random trees and the classical graphs. The reason for this exponential decay is clear. The number of connections of a node depends on its age. The 'old' nodes certainly get more links than the 'young' ones. For example, the mean degree of the oldest node—the root—equals $1 + \frac{1}{2} + \frac{1}{3} + \ldots + \frac{1}{t} \cong \ln t$ in the network of $t+1$ nodes, while the youngest node has a single link.[1] This results in a wider degree distribution than a Poisson one.

[1] More generally, the mean degree of a node born at time u varies as $\ln(t/u)+1$.

There is another important distribution function—the distribution $p(q, u, t)$ of the number of connections q of a node born at a given time $u < t$. To measure this distribution we must grow many members of the ensemble by repeating the growth process many times. One can show that the resulting distribution is Poissonian. It is known that if the first moment of a Poisson distribution is large, then this distribution is relatively narrow, its width is about the square root of the first moment. For example, the mean degree of the root is $\ln t$, and possible fluctuations from this value are only about $\sqrt{\ln t} \ll \ln t$. To be clear, if we grow a single random recursive tree and measure the degree of its root, then we obtain the value $\ln t \pm \sqrt{\ln t}$.

One should note that these results for degree distributions are qualitatively valid even if the random recursive network is not a tree, see Fig. 3.6 (a). It is only important that the attachment is uniformly random. There are, however, interesting characteristics, relevant only for

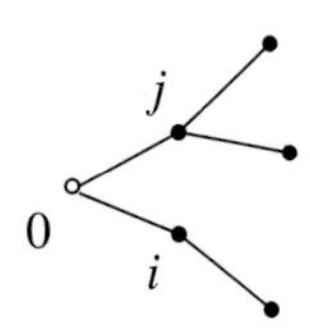

Fig. 7.1 In this recursive tree, node i has one descendant, node j has two, and the root 0 has five. The root has two branches of sizes 2 and 3.

recursive trees. We mention here only two of these. The first is the distribution $\mathcal{P}_d(s)$ of the number of descendants of a node in this tree, see Fig. 7.1. In 2001 Krapivsky and Redner found that this distribution decays very slowly, as s^{-2} [115]. Moreover, it was shown that this law for $\mathcal{P}_d(s)$ is practically general for recursive trees. The uniform attachment was not necessary. For branches of the root, even more slowly decaying distribution of branch sizes was found [92]. The mean number of branches with s nodes is exactly $1/s$. That is, in the random recursive tree, on average, the root has one leaf ($s=1$). Thus, for some distributions, power laws can be found even in the random recursive trees. Researchers, however, were more interested in degree distributions.

7.2 The Barabási–Albert model

In 1999 Barabási and Albert published a truly timely paper addressing the problem of heavy-tailed degree distributions [14]. In this celebrated work they presented a very simple growth model—actually, a concept—pretending to explain the scale-free architectures of real-world networks. The idea was to grow a network, preferentially adding new connections to already highly connected nodes, that is, to exploit the 'rich get richer' effect. Barabási and Albert named this process *preferential attachment*—preferential attachment of new nodes to existing nodes of higher degrees [4]. It is obvious that preferential attachment should increase the number of hubs. The question is: how can it produce scale-free networks?

The Barabási–Albert model is an undirected recursive network, Fig. 3.6 (b), defined as follows. The growth starts from some initial configuration of nodes and links which is actually not important for the structure of the network when it is already large. At each time step,

(i) add a new node to the network and

(ii) attach it to $m \geq 1$ preferentially chosen nodes. Each of these nodes is chosen with a probability proportional to its degree—proportional preference.

Since each new node has m links, the mean degree of a node in this network is $2m$. In particular, $m=1$ corresponds to a tree. The tail of the degree distribution, however, turns out to be independent of m. It is important that the probability of selecting a node of degree q in this process is $q/(2mt)$. When this network is already large, of $t \to \infty$ nodes, its degree distribution can easily be found.[2] The result is a stationary power-law distribution, $P(q) \propto q^{-\gamma}$ with exponent $\gamma = 3$. Of course, this is a very particular power law, but the Barabási–Albert model only demonstrates the principal possibility of generating a scale-free distribution by preferential attachment. Note that the model assumes proportional preferential attachment only for simplicity. In the next sections we will show how to obtain power laws with arbitrary exponents and more general heavy-tailed distributions.

[2] Consider the mean number $\langle N(q,t)\rangle$ of nodes of degree q in a network of size t. For brevity, set $m=1$. Then, for two consecutive moments,

$$\langle N(q,t{+}1)\rangle = \langle N(q,t)\rangle + \frac{(q-1)\langle N(q-1,t) - q\langle N(q,t)\rangle}{2t} + \delta_{q,1}.$$

The second term on the right-hand side is due to attachments to nodes of degree $q-1$ with probability $(q-1)/(2t)$ and to nodes of degree q with probability $q/(2t)$. The Kronecker symbol $\delta_{q,1}$ describes the increase in the number of nodes of degree 1 due to a new node. The degree distribution is $P(q,t) = \langle N(q,t)\rangle/t$. Insert this into the equation, assume that the resulting distribution is stationary at $t\to\infty$, and finally approximate the discrete difference in degree by a derivative d/dq in the range $q\gg1$. This gives the equation:

$$P(q) = -\frac{1}{2}\frac{d[qP(q)]}{dq}.$$

Its solution is $P(q) \propto q^{-3}$.

One can easily see that for $m > 1$, the Barabási–Albert model and other recursive networks of this kind have a locally tree-like structure if a network is large.[3] So its clustering coefficient approaches zero in the large network limit ($t = N \to \infty$) [40]:

$$C \approx \overline{C} \sim (m-1)\frac{(\ln N)^2}{N} . \tag{7.1}$$

The low clustering, however, is not related to the preferential attachment mechanism. There are many network models with preferential attachment and, simultaneously, with high clustering and many loops. As one could expect, this network is a small world, but, interestingly, there is a marked difference in the small-world effect at $m = 1$ (a tree) and $m > 1$ (a locally tree-like network). The mean internode distances grow with N differently [41]:

$$\overline{\ell}(m=1) \sim \ln N \quad \text{but} \quad \overline{\ell}(m>1) \sim \frac{\ln N}{\ln \ln N} . \tag{7.2}$$

Thus long loops play an important role. They make this network more compact.

[3] There is a very small chance that a new node will be attached to closely separated nodes.

7.3 General preferential attachment

One should note that the general idea of using preferential choice to obtain heavy-tailed distributions emerged long before the boom in complex networks.[4] In 1925 G. Udny Yule published 'A mathematical theory of evolution based on the conclusions of Dr. J. C. Willis' which at first sight was very far from networks [183]. Yule proposed a possible explanation for the earlier observed power-law distribution of the number of species in genera. This distribution is a power law with exponent close to 2. The idea of Yule was ultimately simple. It is known that new species emerge due to mutations. Yule supposed that the evolution of a set of genera is a growth process determined by (i) the emergence of new genera with a single species and (ii) the emergence of new species in already existing genera. Two possibilities were taken into account. (1) There is a small chance that mutations create a new genus. (2) Mutations in genera with more species are more frequent and so the probability of the emergence of a new species is greater. The latter dramatically increases the number of large genera. Similarly to the Barabási–Albert model, if the frequency of the emergence of a new species is proportional to the number of species in a genus (a very natural assumption), then this growth leads to a power-law distribution. It is very easy to check that its exponent is indeed 2. Processes of this kind are often called *Yule processes*.

[4] For a brief history of concepts of power-law distributions and relations between various models see [127, 42, 162].

Power laws have been observed in many other distributions. Vilfredo Pareto found this in an income distribution (1897). J. B. Estoup (1916) and George Kingsley Zipf (1932) observed a power-law distribution for the frequency of occurrence of distinct words in a text (exponent close to 2). The distribution of city size (F. Auerbach, 1913) and the distribution of the number of papers by a scientist (Alfred James Lotka, 1926) are also power laws. And these are only a few examples. In 1955 Herbert A. Simon (later Nobel Prize winner, 1978) mathematically developed and presented the ideas of Yule in what is usually called the *Simon model* [163]. Numerous scale-free distributions were interpreted using this model. It is just common sense that the rich attract more wealth, a popular girl attracts more looks, a failure attracts more misfortunes. One may say: 'money attracts money', 'popularity is attractive', and so on. According to Yule and Simon, processes underlying these truisms may result in power-law distributions. In general terms, the mechanism of the emergence of power-laws in these processes is *self-organization*. While growing, systems of this type self-organize into states with scale-free statistics. Sometimes physicists treat these states as 'critical'.

A few 'premature' works of the 1970s and 1980s considering these ideas in the context of networks had no immediate impact.[5] It was the paper of Barabási and Albert that initiated a wave of studies exploiting this mechanism. The Barabási–Albert model only demonstrates the essence of the preferential attachment mechanism. To approach real-world networks, further steps are needed. The first natural step is to find a way to generate scale-free networks with an arbitrary degree distribution exponent and not only 3. For this it is sufficient to change slightly a rule of preferential attachment. Let the probability of attaching to a node of degree q be proportional to a function $f(q)$ which is called a *preference function*. In the Barabási–Albert model, the preference function is q, that is the proportional preference. Let us change in this model just the preference function. Let it now be not proportional but linear, $q + A$. Here $A \equiv am$ plays the role of an 'additional attractiveness', a is a constant number, m is the number of links of each new node. We assume $a > -1$. It turns out that this linear preference leads to exponent

$$\gamma = 3 + a\,, \tag{7.3}$$

which can take values in the range from 2 to infinity [81].[6] Note that the limiting case $a \to \infty$ corresponds to a random recursive graph with uniform attachment.

One can obtain the same result as with linear preference, mixing proportional and uniform preferential attachment. With some probability, make connections as in the Barabási–Albert model and with the complementary probability attach to uniformly randomly selected nodes. These straightforward generalizations provide far more realistic networks than the Barabási–Albert model, but they are not more difficult to work with. For numerical simulations, researchers usually need to generate very large networks, say of 10^7 nodes or more. This can be done easily if a graph is recursive and preference is linear.[7] Still, there is a natural question: what is the origin of the proportional or linear attachment? Indeed, preferential attachment explains power laws, but what can explain preferential attachment? Unfortunately, nobody has found a completely satisfactory, general answer up to now, and this is the weakest point of the self-organization (or self-organized criticality) concept for networks.

[5] In bibliometrics, the model of Derek de Solla Price (1976) based on preferential choice explained the statistics of growing directed networks of scientific citations [153]. In graph theory, Szymański defined recursive trees with proportional preferential attachment [174], which actually coincide with the particular case $m = 1$ of the Barabási–Albert model.

[6] Similarly, one can grow scale-free directed networks. Suppose, for example, that in the recursive network which we discussed, all links are directed from new to old nodes. Suppose that the attachment probability is a linear function of in-degrees, $q_{\rm i} + \tilde{a}m$, $\tilde{a} > 0$. Then the exponent of the in-degree distribution is $\gamma = 2 + \tilde{a}$.

[7] For generating these networks, it is inefficient to select nodes by using their degrees. Indeed, for each attachment, we would have to examine all the degrees. Instead, form a list where each node is repeated as many times as its degree (or, if you want, degree minus m). Selecting uniformly an entry from this list, we get proportional preference. To get linear preference, form in addition the relevantly weighted (with weight a) list of all nodes, and choose uniformly from the combination of these two lists. After every attachment, add new entries to these lists.

One of the possible explanations is a process based on uniform selection. In other words, we should arrive at the proportional (or linear) preference using only a uniform choice. Recall that in an arbitrary network the degree distribution of an end node of a uniformly randomly chosen link is $qP(q)/\langle q\rangle$. So, in principle, choosing a link at random and then its end, we get the proportional preference. The problem is that we have to choose from the set of all links, for which we need to know all of them. There is a better option. To explain, let us consider a network of scientific citations and look how references emerge in scientific papers. A researcher either learns about a paper, which he will cite, directly from journals and archives, or he finds a reference to it in another article. The second option results in preferential attachment. Note, however, that choosing one nearest neighbour of a uniformly selected node, we get the proportional preference only if a network is uncorrelated, which is usually not the case. For an arbitrary network, we have to do the following. Choose a node uniformly and select each of its nearest neighbours with some given nonzero probability.[8] It is easy to show that this gives the proportional preference. The reader may check this claim using any small graph. Here, application to the networks of scientific citations is not unique. Actually the same process drives the evolution of genetic networks [143]. In cellular biology, this process is called *duplication–divergence*. In protein–protein interaction networks,[9] new proteins are born as identical copies of the original ones (duplication), but afterwards new proteins lose some of their functions (divergence), see Fig. 7.2. (In addition, they acquire new functions but this can be treated separately.) In network terms, duplication–divergence means precisely the connection of a new node to a fraction of the nearest neighbours of a selected node. Thus we again arrive at a proportional preferential attachment.

[8] Note that you have a chance to select simultaneously several nearest neighbours, or zero. So this process effectively generates short loops.

[9] Nodes in these networks are proteins, while links indicate functional and other associations between protein pairs.

The preferential attachment mechanism can also be applied to equilibrium networks, but in this section our focus is only on recursive graphs. Nodes in these networks are distinguished (labelled) according to their birth times. It is obvious that older nodes are better connected. Indeed, the mean degree of a node born at time u was found to be $\langle q(u,t)\rangle \sim (u/t)^{-\beta}$ [81]. Here exponent β is related to the exponent of the degree distribution: $\beta = 1/(\gamma - 1)$. Thus we have two related power laws. In comparison with uniform random recursive graphs, the linear preferential attachment sharply changes the statistics of connections of a node born at a given time. Recall that in the model with uniform attachment, fluctuations of $q(u,t)$ are very weak. In contrast to this, due to linear preferential attachment, the degree $q(u,t)$ of node u fluctuates strongly. The fluctuations are of the order of the mean value $\langle q(u,t)\rangle$.

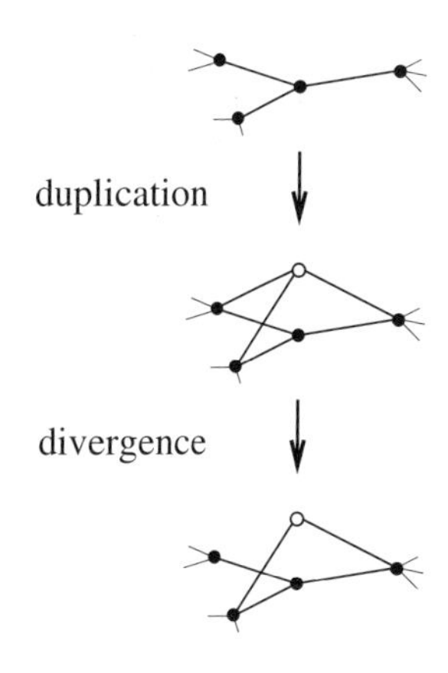

Fig. 7.2 A duplication–divergence event. The white node is a new protein.

Power laws in these recursive networks have been observed not only for degrees and degree distributions. We have already mentioned that in a wide range of recursive trees (with and without preferential attachment) the distribution of the number of ancestors of a node decays as a power law with exponent 2. Similarly, it was found that in all these recursive trees, the betweenness centrality distribution is a power law with the same exponent 2 [97].[10] In these networks—recursive trees with linear preferential attachment—it is even possible to find distributions with exponent less than 2. One example is the size distribution for branches of the root. Let the exponent of the degree distribution of this recursive tree be γ. Then the mean number of branches of size s decays as $s^{-\gamma/(\gamma-1)}$. Exponent γ varies in range between 2 and infinity, so $\gamma/(\gamma-1)$ takes values in the range from 2 to 1.

[10] The coincidence of these exponents is not occasional. One can show that these two distributions are closely related.

The growth of a network starts from some initial configuration of nodes and links. What is its role? Can it influence the evolution of a network, in particular, the form of its degree distribution? In principle, the answer depends on the network model, but for recursive networks, the situation is clear. Figure 7.3 shows a typical degree distribution of a finite scale-free recursive network. As is natural, a power law describes the distribution only in the intermediate region, between low and high degrees. The cut-offs of the degree distributions in these networks increase with size N: $q_{\text{cut}} \sim N^{1/(\gamma-1)}$. Note a bulge near the cut-off in the figure which is visible at any network size. This hump is just the trace of an initial condition—the first nodes will always remain best connected. By changing an initial configuration, we can increase, or diminish, or sometimes even wipe out the hump.

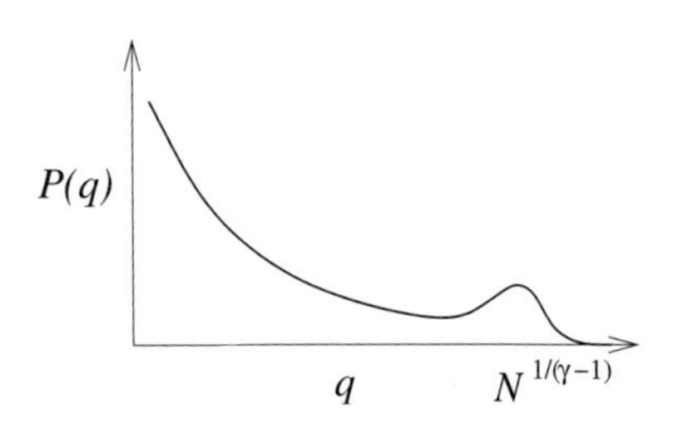

Fig. 7.3 Typical degree distribution of a finite scale-free recursive network.

We have provided some reasons why linear preference can emerge, but, in general, a preference function can be nonlinear. Suppose, for example, that a preference function is a power law, q^{ϕ}, with an arbitrary positive exponent ϕ. Can it produce scale-free distributions? Pavel Krapivsky and Sid Redner showed this is possible only if $\phi = 1$ [114,115]. When the function grows slower than q, that is $\phi < 1$, the resulting degree distribution decays slower than exponentially but faster than a power law. The distribution has a so-called stretched exponential form: $P(q) \propto \exp(-Cq^{1-\phi})$, where the value of C is determined by exponent ϕ. If, on the other hand, the preference function grows faster than q, that is $\phi > 1$, we again fail to get a scale-free distribution. In this case, all links in the growing network become seized by a tiny fraction of the oldest nodes or even by a single node. Thus, the preferential attachment mechanism generates power-law distributions only if preference is linear, which is a serious restriction.

In this lecture we have discussed only a few basic networks demonstrating the essence of the preferential attachment mechanism. Plenty of more detailed and realistic network models based on this mechanism have been considered. We stress that apart from preferential attachment, there are other ways, maybe not so extensively explored in networks, which allow one to generate power-law distributions. We will touch upon some of these alternative ways in the following lectures. Here we will mention only one of the alternative processes. Suppose that the evolution of a network incorporates the merging of nodes. Say, at each time step, a pair of some nodes merge together, and so the corresponding degrees are added together. It turns out that this merging can result in a variety of scale-free degree distributions [109].

7.4 Condensation phenomena

Discussing preferential attachment, we have focused on growing networks. Note, however, that growing systems are only a particular case of nonequilibrium ones. One can easily construct a nonequilibrium network with a constant number of nodes and links.[11] Main qualitative conclusions for the particular case of growing networks with preferential attachment are valid for nonequilibrium networks. On the other hand, it is more difficult to arrive at power-law degree distributions for equilibrium networks by using preferential attachment. It is time to discuss these networks in more detail.

[11] Consider, for example, the following process generating a non-equilibrium network of a fixed size. At each time step, reattach all the emanating links from a randomly chosen node to some other specially selected nodes. Clearly, the statistics of connections of a node in this network depends on the time elapsed after the reattachment.

Let us implement preferential attachment to an equilibrium net. For example, at each time step, choose a link uniformly from the set of all links and reattach one of its ends to a preferentially chosen node. As is natural, the resulting network architecture depends on a preference function. As in growing networks, we must use a linear preference to arrive at scale-free architecture. For equilibrium networks, this is, however, not sufficient. It turns out that the organization of connections is essentially determined by the mean degree of nodes in a network.

(i) If the mean degree is below some critical value, which is determined by the form of a preference function, then the degree distribution is rapidly decaying.

(ii) At the critical value, the degree distribution is power-law.

(iii) Above the critical value, one of the nodes attracts a finite fraction of all connections in the network. The connections of the rest of the nodes are described by a power-law degree distribution with the same exponent as at the critical point, see Fig. 7.4.

This condensation of links frequently occurs in the models of complex networks. As far as we know, condensation has never been observed in real-world networks. The difficulty is that for observation, a network must be very large. Otherwise the hub—a node attracting a finite fraction of connections—cannot be reliably distinguished from other strongly connected nodes.

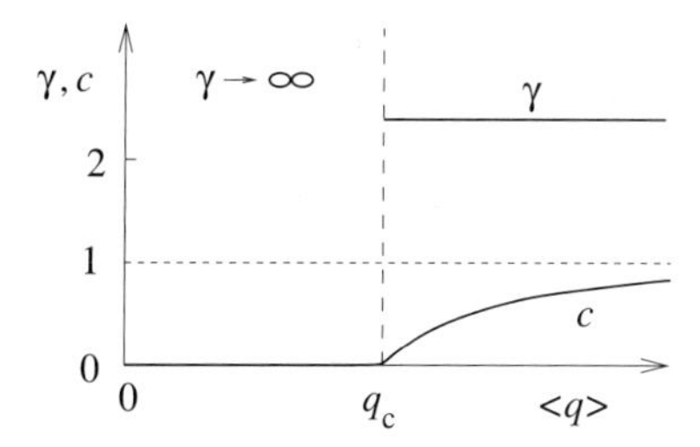

Fig. 7.4 Condensation of links in an equilibrium network. Fraction c of links attached to a single node and the exponent γ of the degree distribution of the rest of the nodes versus the mean degree $\langle q \rangle$. Below the critical value q_c, the degree distribution decays exponentially.

The same equilibrium network, with condensation of links and a power-law degree distribution, can be constructed by other means, without using the preferential attachment. Simply give higher weights to those statistical ensemble members that have many strongly connected nodes [45]. More generally, one can introduce the 'energy' of a graph—a quantity which is determined by some structural characteristics (for example, the total number of triangles—transitivity—in this graph) and then implement Boltzmann statistics to the ensemble of the graph. So here a given form of the 'energy' functional defines a network. After David Strauss [170], random networks constructed in this, natural for statistical mechanics way, are usually called *exponential random graphs*. Importantly, these networks easily develop condensation: of links, triangles, and other structural units, depending on the definition of 'energy'. Models of this kind are usually simulated based on the so-called *Metropolis algorithm.*[12] Usually these simulations start from a homogeneously random configuration, and a final condensed state is approached only after a series of steps. In small networks, this transition is rapid, but, remarkably, it takes an astronomical time even if a network is moderately large [46].

[12] This is one of the standard algorithms for physics and optimization problems [102]. In application to spin systems ($S_i = \pm 1$), the algorithm works as follows. At each time step, select at random a spin, say, in state S_i and compare the energies of two configurations—with S_i and with $-S_i$. If the latter has a lower energy than the former, then flip the spin: $S_i \to -S_i$. Otherwise pass to the next step, and so on until the system approaches an equilibrium state.

Condensation also occurs in growing networks. Here we consider one of the possibilities. Let us return to recursive networks with a linear attachment, that is, the preference function is $f(q) = hq + a$. We assumed previously that the parameters h and a of the preference do not depend on a node.[13] In this respect, all nodes were assumed equal. In real-world networks, this is apparently not the case—nodes are very different and have different potentials for growth. Compare Google and your home page. To model this diversity, Ginestra Bianconi and László Barabási assumed that the parameters a and h vary from node to node [28]. A 'young' node with large a and/or h has a good chance to quickly attach many links and become more connected than older nodes with low parameters a and h. If fluctuations of the preference parameters are low, the effect of this heterogeneity is not very strong—the exponent of a degree distribution changes. If, however, the fluctuations of a and, especially, h, are sufficiently strong, the heterogeneity changes qualitatively the organization of a network. A finite fraction of connections become attracted by one of the nodes [29]. The degree distribution for the rest of the nodes is scale-free. For simplicity, assume that only one node in a network is distinct from the other, namely that this node has $h > 1$, while others have $h = 1$. Figure 7.5 for this particular case shows how the fraction c of condensed links and the exponent γ of the degree distribution depend on h (compare with Fig. 7.4 for condensation in equilibrium networks).

[13] In this situation, we could use $f(q) = q + a/h$.

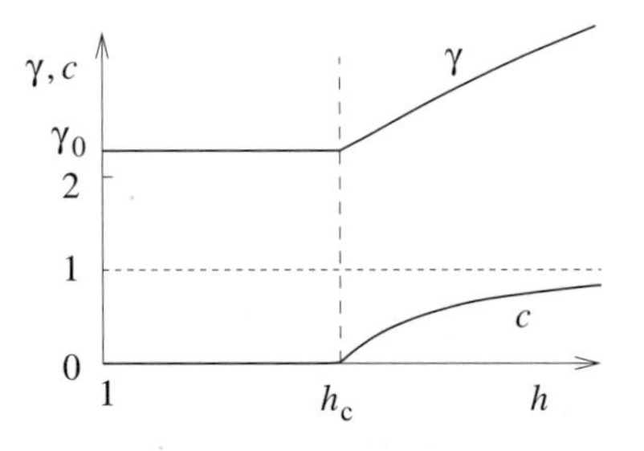

Fig. 7.5 Condensation of links in a growing network in which only one node has the preference function parameter $h > 1$ while others have $h = 1$, see book [80]. Fraction c of condensed links and the exponent γ of the degree distribution versus h. In the homogeneous network ($h = 1$ for all nodes) the value of the degree distribution exponent is assumed to be $\gamma_0 > 2$.

7.5 Accelerated growth

We have already mentioned that in many real-world networks the number of links grows faster the number of nodes, so that the mean degree of a node is a growing function of the network size [120]. This nonlinear, or, one can say, 'accelerated', growth has numerous consequences [78, 79].

We indicated that, for example, the diameter of a growing network of this kind may be constant or even decrease as the number of nodes increases. The models of networks that we have considered up to now do not show this accelerated growth. It is not very difficult, however, to incorporate this feature into network models with preferential attachment. Suppose, for example, that in a random recursive network, the number of connections of a new node $m(N)$ is a growing function of N. Then the mean degree of a node, $\langle q\rangle(N) = 2m(N)$, is also growing. With the acceleration, a network can still be scale-free, if the preferential attachment is linear. There are some differences compared to previously discussed degree distributions. Without the acceleration, a scale-free degree distribution is stationary in the limit $N \to \infty$, and its exponent

γ cannot be smaller than 2, otherwise the first moment of a degree distribution would diverge. This is the case in most of the studied models with preferential attachment. If the growth is accelerated, both these restrictions are removed: degree distributions may be non-stationary, and γ may occur between 1 and 2. Low values of this exponent have indeed been observed in a number of real networks of this kind. For example, γ is below 2 in the degree distributions of large software maps [178].

7.6 The BKT transition

In each of the growing networks discussed above, all nodes were within a single connected component. More generally, growing networks, similarly to equilibrium ones, can consist of a rich set of connected components. How are they organized in a growing network and what is their difference from equilibrium networks?

For demonstration, consider the following network [49] with numerous connected components. The growth process consists of two channels: bare nodes are added at unit rate and links connecting uniformly chosen nodes are added at rate b. The rate for links, b, plays the role of a control parameter which determines the structure of the network. Clearly, the mean degree of a node in this network is $\langle q \rangle = 2b$. One can easily show that this growing random graph has an exponentially decaying degree distribution. If $b = 0$, the network is a set of bare nodes. With increasing b, the network contains bigger and bigger components. If the parameter b is greater than some critical value b_c, then the growing network has a giant connected component consisting of a finite fraction S of all nodes. The remarkable finding was that this giant component emerges in a drastically different way from that in equilibrium networks [49]. Namely, near the birth point, the relative size of a giant component is

$$S \propto \exp\left(-\frac{D}{\sqrt{b - b_c}}\right), \tag{7.4}$$

where D is a positive constant. So, any derivative $d^n S(q)/dq^n$ ($n = 0, 1, 2, \ldots$) is zero at the critical point. In standard classification, this is an infinite-order transition. Compare this singular function of b to the critical singularities of S in equilibrium networks, which we listed in the preceding lecture. In condensed matter physics, this so-called *Berezinskii–Kosterlitz–Thouless* (BKT) singularity is considered to be rather exotic. For theoretical physicists, it was surprising to observe this anomaly in small worlds. Similar critical features were later observed in many other growing networks, including scale-free [82].

There is another marked difference. In the absence of a giant component in these networks, the size distributions $\mathcal{P}(s)$ of finite components decay in a power-law fashion. In physics, a power-law decay often indicates a critical state. So in these networks, the entire phase without a giant component can be called 'critical', and not only its birth point.

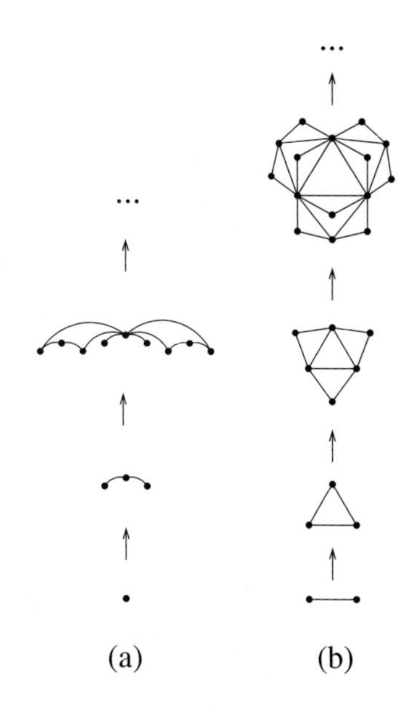

Fig. 7.6 Two of the constructions of scale-free deterministic graphs [16, 74].

7.7 Deterministic graphs

In this lecture we have mostly discussed growing networks distinguished by three features: (i) the small-world feature, (ii) a scale-free architecture, (iii) randomness. It is possible, however to build more simple non-random scale-free small worlds, in other words, deterministic scale-free graphs. Two examples, shown in Fig. 7.6, explain this construction. The diameters of these graphs grow logarithmically ($\ln N$) with their size, typically for small worlds but in sharp contrast with fractals (see Fig. 1.5). The degrees of deterministic graphs take a discrete set of values in contrast to random networks. On the other hand, the envelope of the occurrence frequency of these degrees is often power-law, in particular for the graphs in Fig. 7.6. In that sense, these graphs are scale-free. Surprisingly, many of their properties are very close to those of random scale-free networks. So these toy models are closer to real-world networks that one would expect.

Correlations in networks

8

In Lecture 5 we considered network models without structural correlations. However, real-world networks are correlated. In this lecture we discuss the simplest correlations in complex networks and explain their nature.

8.1 Degree–degree correlations

Among a wide spectrum of various structural correlations in networks, the simplest are correlations between degrees of nearest-neighbouring nodes. These correlations are described by the joint distribution of these degrees, $P(q, q')$. Usually, it is called a *degree–degree distribution.* This is the probability that a randomly chosen link connects nodes of degree q and q'.[1] Recall that in an uncorrelated network, this degree–degree distribution is determined by the degree distribution $P(q)$ of the network, namely, $P(q, q') = qP(q)q'P(q')/\langle q\rangle^2$. This is not true, if correlations are present. In an arbitrary network, the degree distribution of an end node for a uniformly chosen link is $qP(q)/\langle q\rangle$. So $P(q)$ is expressed in terms of a degree–degree distribution:

$$\sum_{q'} P(q, q') = \frac{qP(q)}{\langle q\rangle} . \tag{8.1}$$

[1] How can we obtain the degree–degree distribution from empirical data? Let the number of links in a graph be L. Find the number of links $L(q, q')$ connecting nodes of degree q and q'. If $q \neq q'$, then $P(q, q') = L(q, q')/2L$, and if $q = q'$, then $P(q, q) = L(q, q)/L$.

Instead of the joint degree–degree distribution, equivalently one can use the conditional probability $P(q|q')$ that if one end node of a link is of degree q', then its second end node is of degree q.[2]

[2] One can see that

$$P(q|q') = \frac{P(q, q')}{\sum_q P(q, q')} = \overline{q}\frac{P(q, q')}{q'P(q')} .$$

The symmetry $P(q, q') = P(q', q)$ immediately leads to the following symmetry relation for the conditional probability [36]:

$$qP(q'|q)P(q) = q'P(q|q')P(q') .$$

Let us take one step beyond uncorrelated networks defining an equilibrium network model with only correlations between degrees of nearest neighbours. In the spirit of the configuration model, this is the maximally random network with a given joint degree–degree distribution $P(q, q')$. Thus we have a hierarchy of equilibrium network models.

(i) The classical random graphs are the lowest level of the hierarchy. These are maximally random networks under a single constraint: $\langle q\rangle$ is fixed.

(ii) The uncorrelated networks are the second level. These are maximally random networks with a given degree distribution $P(q)$. As a consequence of this constraint, a mean degree $\langle q\rangle$ is also fixed.

(iii) Maximally random networks with a given $P(q, q')$ are the third level. This constraint ensures that $P(q)$ is also fixed, and then $\langle q\rangle$ is fixed as well.

One can continue this hierarchy by imposing more and more rigid constraints which leads to more complex multi-degree correlations. Unfortunately, already, the third level of hierarchy is sufficiently difficult for analysis.

8.2 How to measure correlations

The complete information about degree–degree correlations in a real-world network can be obtained only by measuring its joint degree–degree distribution $P(q, q')$ or $P(q|q')$. Alas, this is practically impossible in sparse networks with heavy-tailed degree distributions. Suppose the number of links in a network is $L \sim N$ and the cut-off of the degree distribution is at q_{cut}. To arrive at a usable $P(q, q')$ with sufficiently small fluctuations, we must have a large number $L(q, q')$ of links connecting nodes of degrees q and q', for each pair of degrees. This is impossible if L is less or of the order of q_{cut}^2, which is the case if a network is sparse and q_{cut} is large. To avoid strong fluctuations in empirical data, researchers have to use a less informative though convenient characteristic, namely the average degree of the nearest neighbour of a node as a function of the degree of this node [148]:

$$\overline{q}_{\text{nn}}(q) = \sum_{q'} q' P(q'|q) \,. \tag{8.2}$$

We mentioned this quantity while discussing real-world networks. This dependence relates the degree of a node and the average degree of its nearest neighbours and allows one to find how nodes of different degrees are interconnected. If a network is uncorrelated, $\overline{q}_{\text{nn}}(q) = \langle q^2 \rangle / \langle q \rangle$, that is a constant. In correlated networks, dependences $\overline{q}_{\text{nn}}(q)$ can be very diverse in shape. There is some restriction. It turns out that the second moment of the degree distribution and $\overline{q}_{\text{nn}}(q)$ are related [37].[3] Namely,

$$\langle q^2 \rangle = \sum_{q'} q' P(q') \overline{q}_{\text{nn}}(q') \,. \tag{8.3}$$

[3] This relation directly follows from the definition of $\overline{q}_{\text{nn}}(q)$.

If, in particular, the second moment of the degree distribution diverges, $\overline{q}_{\text{nn}}(q)$ cannot be arbitrary. Its form must guarantee the divergence of the right-hand side of relation (8.3). If so, then what are the possible dependencies $\overline{q}_{\text{nn}}(q)$?

8.3 Assortative and disassortative mixing

Sociologists usually characterize degree–degree correlations by an even more rough quantity than $\overline{q}_{\text{nn}}(q)$. This is *the Pearson correlation coefficient* defined as

$$r = \frac{\langle qq' \rangle_l - \langle q \rangle_l \langle q' \rangle_l}{\langle q^2 \rangle_l - \langle q \rangle_l^2} \,. \tag{8.4}$$

Here q and q' are the degrees of the end nodes of a link, and $\langle \; \rangle_l$ denotes the average over all links in a network.[4] The Pearson coefficient is a

[4] Equivalently, one can substitute $q-1$ and $q'-1$ in this definition for q and q'. $\langle qq' \rangle_l$ is the average product of the degrees of the end nodes of a link. $\langle q \rangle_l$ is the average degree of an end node of a link. As we explained, $\langle q \rangle_l = \langle q^2 \rangle / \langle q \rangle$ for an arbitrary network. Similarly, the average square of the degree of an end node of a link, $\langle q^2 \rangle_l$, equals $\langle q^3 \rangle / \langle q \rangle$. Expression (8.4) is not very convenient for the computation of r from empirical data. See [134] for a more convenient form of this definition.

standard pair correlation function $\langle qq' \rangle_l - \langle q \rangle_l \langle q' \rangle_l$ normalized in such a way that r is in the range between -1 and 1. Clearly, if a network is uncorrelated, then the Pearson coefficient is zero. If, on average, strongly connected nodes have strongly connected neighbours, then $r > 0$. These correlations are called *assortative*, and this situation is referred to as *assortative mixing* [134]. In the opposite situation, on average, a weakly connected node has a strongly connected neighbour and vice versa. This is called *disassortative mixing*, and for these correlations, the Pearson coefficient is negative. Figure 8.1 shows examples of networks with ultimately strong assortative and disassortative mixing, $r = 1$ and -1, respectively.

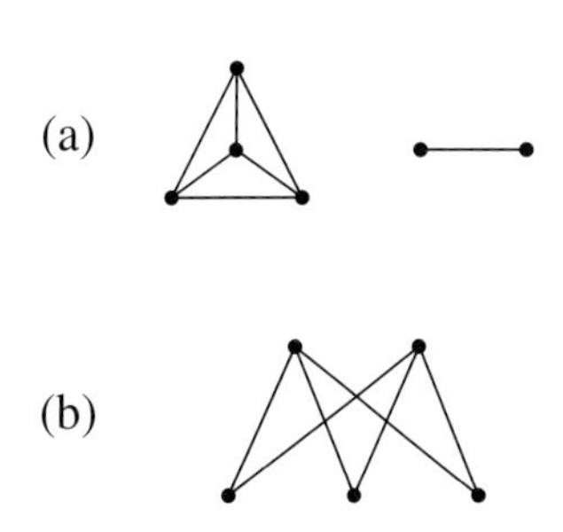

Fig. 8.1 Two small networks with ultimately assortative (a) and disassortative (b) mixing. In network (a), only nodes of equal degree are interlinked. In network (b), only nodes of different degrees are interlinked. Check that r equals 1 and -1 for (a) and (b), respectively.

We can distinguish between assortative and disassortative correlations in another way by inspecting the dependence $\overline{q}_{\rm nn}(q)$, see Fig. 8.2 (a). Clearly, if this curve is monotonously growing, then the degree–degree correlations are assortative, if it is monotonously decreasing, then the correlations are disassortative. The problem is that the curve $\overline{q}_{\rm nn}(q)$ is often not monotonous as in Figs. 8.2 (b) and 4.4, and then clear distinction between the two kinds of correlations is impossible. Furthermore, sometimes, the Pearson coefficient is itself confusing. To see this, we rewrite expression (8.5) using $\overline{q}_{\rm nn}(q)$. This gives

$$r = \frac{\langle q \rangle \sum_q q^2 \overline{q}_{\rm nn}(q) P(q) - \langle q^2 \rangle^2}{\langle q \rangle \langle q^3 \rangle - \langle q^2 \rangle^2} . \tag{8.5}$$

Note the third moment of the degree distribution, $\langle q^3 \rangle$, in the denominator. In infinite scale-free networks, $\langle q^3 \rangle$ diverges if exponent γ is less or equal to 4. So if the numerator is finite, we get zero Pearson's coefficient in an infinite network. It turns out that this is the case only if degree distribution exponent γ is between 3 and 4. In this situation for finite networks, r strongly depends on network size. So, it is practically impossible to compare the values of the Pearson coefficient for different networks and, for example, determine which one of them is 'more assortative' [73]. Only if exponent γ is outside of this $(3, 4)$ interval, is r finite in infinite networks and can it be used for the characterization of correlations.[5] Furthermore, zero Pearson's coefficient does not guarantee the absence of degree–degree correlations. In principle, some non-monotonous $\overline{q}_{\rm nn}(q)$ substituted in relation (8.5), may give $r = 0$.

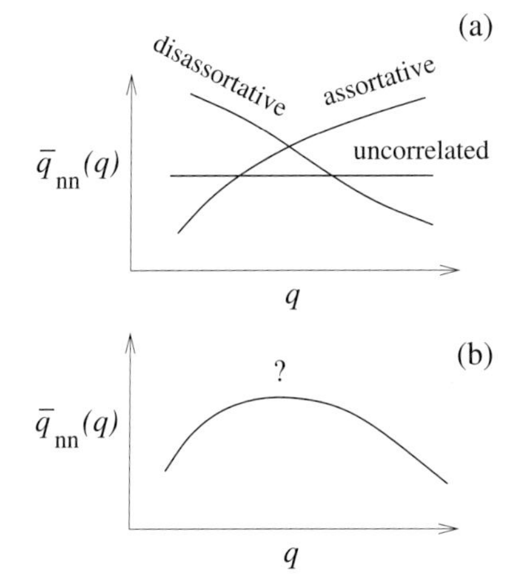

Fig. 8.2 (a) Typical curves $\overline{q}_{\rm nn}(q)$ for various kinds of degree–degree correlations. (b) It is impossible to conclude whether the mixing is assortative or disassortative by inspecting this non-monotonous dependence.

These drawbacks of r seriously hamper comparison of empirical data for different networks with heavy-tailed degree distributions. Nonetheless, the Pearson correlation coefficient is always listed among main network characteristics, providing information at least about the type of degree–degree correlations. For example, practically all empirical data on social and collaboration networks indicate strong assortative correlations (see a large survey of empirical data on networks [66]). That is, a sociable person typically has sociable friends, while (very few) friends of an introvert are also unsociable. On the other hand, many technological networks demonstrate disassortative mixing. In particular, the Internet on the Autonomous Systems level has these kinds of correlations [150]. However, the network of routers in the Internet does not have such clear

[5] If $\gamma < 3$, in infinite networks, both the numerator and denominator on the right-hand side of eqn (8.5) diverge, and these divergences compensate each other resulting in a finite r. The divergence of the numerator is explained by relation (8.5). It was found that $\overline{q}_{\rm nn}(q)$ is strongly size-dependent in this range of exponent γ [37].

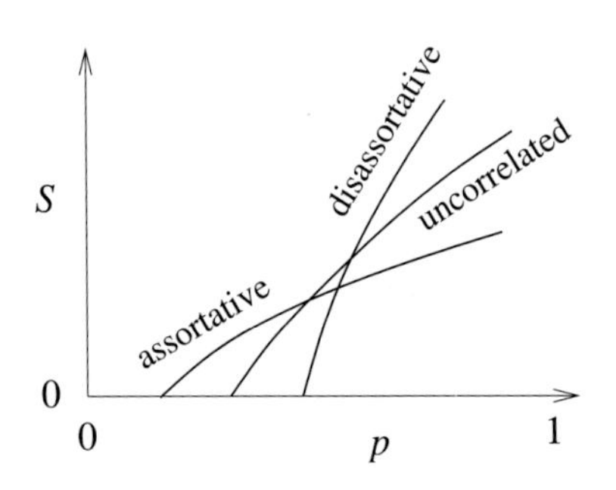

Fig. 8.3 The effect of degree–degree correlations on the emergence of a giant connected component. The size of a giant component versus the retained fraction p of nodes after damaging the network.

degree–degree correlations, see Fig. 4.4. The WWW and the network of protein interactions have disassortative degree–degree correlations.

What are the consequences of degree–degree correlations? To be specific, let us discuss how degree–degree correlations influence global organization of a network, namely a giant connected component. Clearly, we should compare correlated networks with uncorrelated ones having identical degree distributions. Consider, for example, the percolation problem. That is, remove a fraction $(1-p)$ of uniformly chosen nodes and inspect how this affects a giant connected component. We assume that the degree–degree correlations are sufficiently weak, so that they can only moderately change the organization of a network. In particular, these weak correlations cannot modify the ultraresilience criterion obtained for uncorrelated networks. If the second moment of a degree distribution is infinite in an infinite correlated network (which corresponds to exponent $\gamma \leq 3$), then the percolation threshold is still $p_c = 0$. The effect of correlations on the percolation threshold is visible if a degree distribution decays sufficiently rapidly, $\gamma > 3$. In this situation, assortative correlations diminish p_c from its value for an uncorrelated network, that is they make the network more resilient against random damage [134]. Disassortative mixing results in the opposite effect.

At first sight, enhancement of resilience due to assortative correlations is quite obvious. Indeed, thanks to these correlations, nodes of high degrees are better interconnected with each other than in uncorrelated networks. This strongly interconnected core part of a network belongs to the giant connected component (if it exists, of course). To eliminate the giant component, one has to split this strongly interconnected core, which is not an easy task. Nonetheless, the effect of correlations is not that simple. Consider the influence of correlations on a full curve—size of a giant component versus p. Figure 8.3 shows that assortative correlations indeed decrease p_c, making it harder to eliminate a giant component, but this is not all. Surprisingly, these correlations may also suppress the size of a giant component [134]. Thus, these two contrasting effects strangely coexist with each other.

8.4 Why are networks correlated?

Real-world networks are correlated. Even if empirical data indicate the absence of correlations in some real-world networks, this usually only means that correlations are weak or that an unsuitable or too crude quantity (e.g. r) was studied. Importantly, all growing network models are correlated. In the growth process, the presence of links between nodes depends on ages and degrees of these nodes, and there is an asymmetry between younger and older nodes. So, correlations are practically inevitable. The type of correlations depends on the model.

Degree–degree correlations are present even in uniform random recursive graphs. Consider more general, scale-free random recursive graphs which are characterized by a degree distribution exponent γ. In these networks, if $\gamma < 3$, then, asymptotically,

$$\overline{q}_{\rm nn}(q) \sim q^{-(3-\gamma)}, \tag{8.6}$$

which is typical for disassortative mixing [115]. If $\gamma > 3$, the slope of $\overline{q}_{\rm nn}(q)$ is opposite, which indicates assortative mixing. The Barabási–Albert model is unique in the sense that at $\gamma = 3$, mixing changes from disassortative to assortative.[6] This is one of the reasons to use less special models with linear attachment and not the Barabási–Albert model.

[6] For $\gamma \geq 3$, $\overline{q}_{\rm nn}(q)$ changes logarithmically slowly [20]. Interestingly, although these networks are correlated at any γ, their Pearson coefficient is zero when a network is infinite [73].

In some equilibrium networks, degree–degree correlations are also inevitable. It turns out that additional structural constraints lead to correlations [123]. In particular, the banning of multiple links is a constraint of this kind. Let us discuss a uniformly random graph with a given degree distribution, but in contrast to the configuration model, now we forbid multiple links and loops of length 1. In this constrained model, the form of the degree distribution is particularly significant. In practice, this network may be built by using the following randomization algorithm [123]. (i) Create an arbitrary graph of a given size, with a given degree sequence, and without multiple connections. (ii) Rewire a pair of randomly chosen links in a way shown in Fig. 8.4. This retains the degrees of all nodes and does not create multiple links. (iii) Repeat (ii) until the network relaxes to an equilibrium state.[7]

[7] Here we only outline the idea of the algorithm.

A difference between the configuration and the constraint models exists only if a network has hubs of degrees of the order of $\sqrt{N}$ or greater.[8] In the configuration model, these hubs have a good chance of being multi-linked and of having 1-loops. Nodes with such high degrees exist only if the degree distribution slowly decays. Exponent γ must be smaller than 3. Furthermore, the cut-off of the degree distribution must grow more rapidly than $\sqrt{N}$. If we forbid multiple links, the excessive links between multiply connected nodes should be rewired. This (i) reduces the probability that high-degree nodes are interlinked, and (ii) increases the probability that the highly connected nodes are linked to weakly connected ones. Consequently, this 'constrained' network has disassortative degree–degree correlations. This also leads to a rough estimate $q_{\rm cut} \sim N^{1/2}$ for an uncorrelated network without multiple links, if exponent $\gamma \leq 3$. There cannot be nodes with higher degrees in networks of this kind.

[8] For simplicity, estimate the mean number of 1-loops at a node of degree $q \gg 1$ in the configuration model. Use the construction shown in Fig. 5.1. Each of the q stubs joins one of the remaining stubs at this node with probability $(q-1)/\langle q \rangle N$. This gives, on average, about $q^2/\langle q \rangle N$ 1-loops at this node.

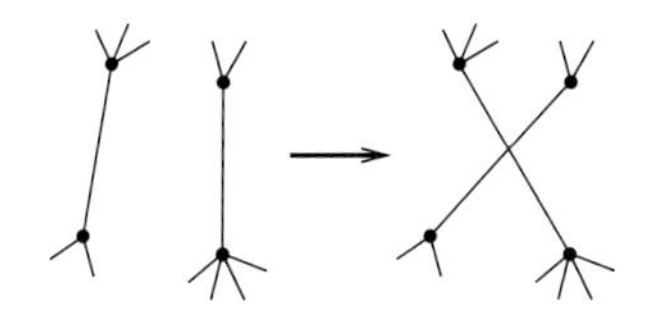

Fig. 8.4 Rewiring operation for the randomization algorithm [123]. This rewiring preserves degrees of nodes and does not create multiple links.

By similar reasoning, strongly connected nodes (with degrees of the order of or larger than $N^{1/2}$) are strongly interlinked between each other. A network is sparse, but its subgraph based on these 'rich' nodes is dense. As was mentioned, this is called a rich-club phenomenon [185]. Remarkably, the rich-club phenomenon is possible in a wide range of networks: correlated and uncorrelated, with and without multiple links. This phenomenon was observed in many real-world networks with slowly decaying degree distributions [63]. For example, the tier-1 Autonomous Systems in the Internet form a rich club.

Although the rich club is relatively small, it plays an important role in a network. It belongs to a giant connected component (if it exists) and form its 'core'. This part of a network is vitally important for its resilience against random damage. Indeed, the rich-club phenomenon is observed for the same slowly decaying degree distributions as the ultra-resilience. Therefore one can significantly improve the robustness of a network by enhancing (or introducing) the rich-club phenomenon. For this, add new links between nodes of high degrees.

Remarkably, correlations may be found even within uncorrelated networks, in particular, within the standard configuration model. Suppose that an uncorrelated network has a finite fraction of dead ends, and so, as we explained earlier, its giant connected component is smaller than the network. Then, counterintuitively, this giant component is correlated.[9]

[9] To see the reason for this, let us assume the opposite, i.e., that the component is itself an uncorrelated network. Since the giant component also apparently has a finite fraction of dead ends, we conclude that this component should contain finite connected components, which is false. Therefore, our assumption was wrong. There is another explanation. An uncorrelated network with dead ends necessarily contains 'dimers'—separated pairs of connected nodes. The giant component of this network has dead ends but no dimers, so it must be correlated.

8.5 Degree correlations and clustering

All networks belonging to the network constructions of Section 8.1 are locally tree-like if a degree distribution decays sufficiently rapidly. In these equilibrium network models, the clustering coefficient vanishes as $1/N$. High concentrations of triangles can be found only in a (small) rich club, if it exists in a network. Suppose that a network has only degree–degree correlations, $P(q, q')$. This distribution determines the clustering coefficient. One can directly generalize formula (5.2) obtained for uncorrelated networks.[10] The key difference is that in correlated networks, the clustering becomes degree dependent. As in real networks, the resulting mean clustering $\overline{C}(q)$ of a node depends on its degree, and so the clustering coefficient does not coincide with the mean clustering, $C \neq \overline{C}$.

[10] See review [77] and references therein.

Clustering in networks, constructed in this way, is only a finite size effect. Mark Newman, however, proposed a direct generalization of the configuration model, which has a finite clustering coefficient even if the network is infinite [137]. His network is made of two simple non-overlapping motifs (frequently repeated subgraphs): triangles (3-cliques) which have no joint sides and single links (2-cliques) which do not belong to any of the triangles. Instead of a given degree sequence, consider a sequence of pairs of numbers. For each node i two numbers are given—the number of triangles to which this node belongs, t_i, and the number of the remaining connections (single links), s_i, see Fig. 8.5. Importantly, all the triangles have no joint sides. So the degree of the node i is $q_i = s_i + 2t_i$. Newman's network is defined as a uniformly random graph with a given $\{s_i, t_i\}$ sequence. Therefore this network is a superposition of two 'configuration models': for single links and for triangles.[11] One can show that, remarkably, this network has degree–degree correlations.

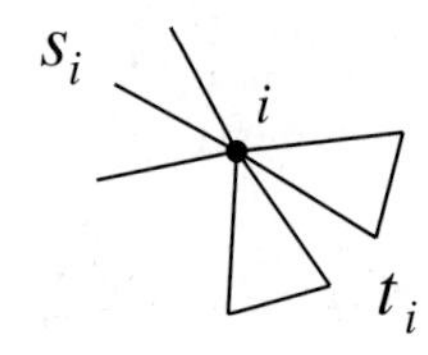

Fig. 8.5 Newman's generalization of the configuration model [137]. For each node i in this network, two numbers are given: the number t_i of triangles to which this node belongs and the number s_i of its rest connections. Note that the triangles have no joint sides.

[11] Check that local clustering in this model cannot be too strong:

$$C_i \leq \frac{q_i/2}{q_i(q_i-1)/2} = \frac{1}{q_i-1}.$$

If we need a clustered network of this kind for numerical simulation purposes, it is more convenient to use a hidden variable construction. Recall our discussion of hidden variables in Lecture 5. In this approach, $\{s_i, t_i\}$, i=1,...,N, is a given sequence of desired numbers of single links and triangles. Using these numbers, connect pairs of nodes (i, j) with probability $p_{ij} \propto s_i s_j$, and interconnect triples of nodes (i, j, k) with probability $p_{ijk} \propto t_i t_j t_k$. Instead of triangles, we could use more complicated motifs—for example, long loops or large cliques, and so on.

Weighted networks

9

In this lecture we consider networks in which links differ from each other. The links are made individual by ascribing a positive number—a *weight*—to each of them. These *weighted networks* enable us to quantitatively represent processes and flows in real-world networks—from various transportation and information nets to social ones.

9.1 The strength of weak ties

In 1973 American sociologist Mark Granovetter published his landmark paper 'The strength of weak ties' [100]. In this remarkable work whose influence has spread far beyond sociology, Granovetter pioneered a quite new understanding of social networks. According to Granovetter, individuals in social networks are connected by 'ties'—weighted links. The 'tie strength' (actually, the link weight in a one-partite weighted network) is defined to be proportional to the frequency (intensity) of social interaction between two individuals [100, 101]. According to strength, we can distinguish 'strong' and 'weak' social ties. In the view of Granovetter, 'our acquaintances (weak ties) are less likely to be socially involved with one another than are our close friends (strong ties). Thus the set of people made up of any individual and his or her acquaintances comprises a low-density network ... whereas the set consisting of the same individual and his or her close friends will be densely knit'. As a result, a social network looks as shown in Fig. 9.1: dense communities of strongly tied close friends are connected together by weak acquaintance ties. In this scheme, 'The weak tie between Ego and his acquaintance ... becomes not merely a trivial acquaintance tie but rather a crucial bridge between the two densely knit clumps of close friends. ... These clumps would not, in fact, be connected to one another at all were it not for the existence of weak ties...'

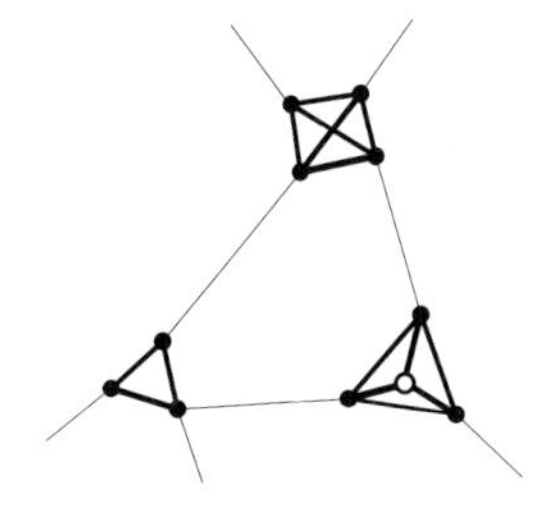

Fig. 9.1 Organization of social networks according to Granovetter. The width of each link represents the strength of the corresponding social tie. Weak ties connect together dense communities of strongly tied individuals. An individual in the centre of the right community (an open dot) has no weak ties and receives any 'external' information only after his close friends.

This organization of social networks has an important consequence also indicated by Granovetter. Ego receives information from his or her close friends through strong ties, while information from other, outer parts of a social network reaches Ego through weak ties. So a lack of weak ties would significantly delay the receipt of information coming from the outer social world. Indeed, in this case, Ego can hear all news only after his or her close friends. This puts Ego in a disadvantaged position, for example, in the labour market (which was a point of interest for Granovetter). Contrastingly, individuals with many weak ties have an apparent advantage in job hunting.

Granovetter assumed this structure of social networks[1] and tested its direct consequence by surveying workers to find who did tell them about their current job. In most cases, the information came from 'not a friend, an acquaintance'. Interestingly, later sociological studies found a number of exceptions from this picture [101]. Recently (2007), researchers found a new possibility of verifying Granovetter's ideas more thoroughly [144, 145]. Large log files stored by cellular network operators contain traces of all mobile phone calls made within these networks. From these records, in particular, one can get the total duration of calls between each pair of customers during a given period. This number characterizes the intensity of social interaction. So the duration of calls can naturally be interpreted as the strength of a given social tie, or, in other words, as the weight of a link in a social network.[2] In this way, these authors constructed a large weighted undirected network of 4.6×10^6 nodes and 7.0×10^6 links. A great majority of nodes in this network (about 85% of the total number) belonged to a giant connected component. The questions were: how does the weight of a link relate to its position in this network? What is the relation between the weight of a link and information flow though this link?

[1] This assumption is often called the *strength-of-weak-ties hypothesis* or simply the *weak ties hypothesis*.

[2] Phone calls have directions, and so, in principle, the log files allow us to construct a directed network. In the studies of Onnela *et al.* [144,145], the directions of phone calls were ignored for the sake of simplicity.

Let us first discuss the former question. To answer this and so to confirm or reject the picture shown in Fig. 9.1 one has to do two things: (i) analyse the structure of the close neighbourhood of a link of a given weight and (ii) reveal the role of links of given weight in the global organization of this network. The closest environment of a tie connecting two individuals i and j is essentially characterized by the relative number of their close friends which are friends of each other. This relative overlap can be characterized by the following expression:

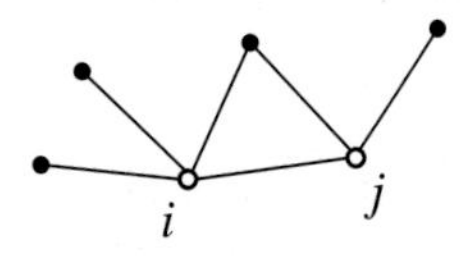

Fig. 9.2 The relative overlap of the common friends of nodes i and j in this configuration equals $O_{ij} = 1/(4-1+3-1-1) = 1/4$. This number differs from the link-clustering coefficient $C_{ij} = 1/(3-1) = 1/2$.

$$O_{ij} = \frac{t_{ij}}{q_i - 1 + q_j - 1 - t_{ij}}, \tag{9.1}$$

where t_{ij} is the number of common friends of nodes i and j, which is also the number of triangles attached to the link; q_i and q_j are the degrees of nodes i and j. Instead of the relative overlap, the link-clustering coefficient

$$C_{ij} = \frac{t_{ij}}{\min(q_i, q_j) - 1} \tag{9.2}$$

can be used (compare with the definition of the clustering coefficient). Here $\min(q_i, q_j) - 1$ is the maximum possible number of common friends for the given q_i and q_j. Figure 9.2 explains these two closely related characteristics. The authors of papers [144, 145] measured the average relative overlap $\overline{O}(w)$ for a link of weight w as a function of this weight. The result was a dependence shown schematically in Fig. 9.3. The monotonously increasing $\overline{O}(w)$ would clearly support the picture of Fig. 9.1—the stronger the tie, the stronger the overlap of friends between two individuals. However, the curve in Fig. 9.3 is non-monotonous. Does this argue against the weak ties hypothesis? It does not. The point is that only a small fraction (about 5%) of links have weights in the range where the curve $\overline{O}(w)$ declines. The remaining links have lower weights,

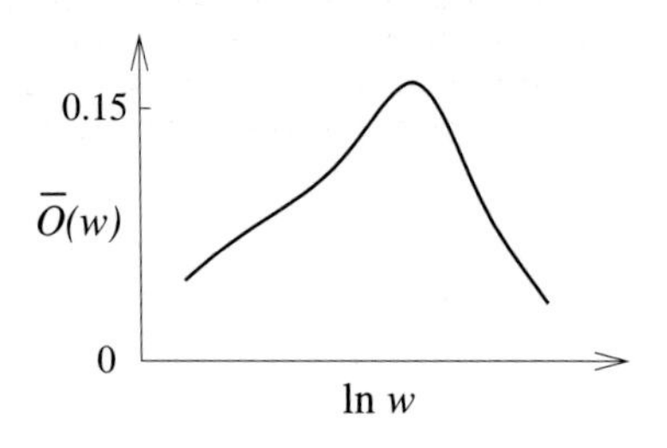

Fig. 9.3 Schematic plot of the dependence of the average overlap $\overline{O}(w)$ for a link of weight w as a function of the logarithm of this weight found in [145]. Only 5% of all links contribute to the declining part of this curve.

being in the region of monotonously increasing dependence. Thus, despite the non-monotonous form, this curve supports Granovetter's hypothesis.

What about the place of weak ties in the global structure of this network? More precisely, how can one verify the hypothesis that the weak ties interconnect dense communities? For this, Onnela and his coauthors used a very standard method of network research. They ranked all of the links in this network according to their weights w and studied how the giant component size changes if links are successively removed based on this ranking: first, starting from high w and, second, starting from low w. Figure 9.4 shows schematically the two resulting curves. The reader can see that the network is disintegrated more rapidly when the weak ties are first broken. This shows that the scheme in Fig. 9.1 is indeed valid. In the same way, instead of S one could study the variations of other network characteristics, global or local. For example, consider the changes in average clustering $\overline{C}$ with f for these two kinds of random damaging—the breaking of links starting from weak and strong ties. This quantity, $\overline{C}$, is, of course, not a global characteristic. It rather characterizes the 'local cliquishness' of a network. It turns out that the removal of links starting from the strong ones diminishes $\overline{C}$ more rapidly than if we delete links starting from the weak ties [145]. This is again consistent with the assumption that strong ties are within dense communities while weak ties connect these communities.

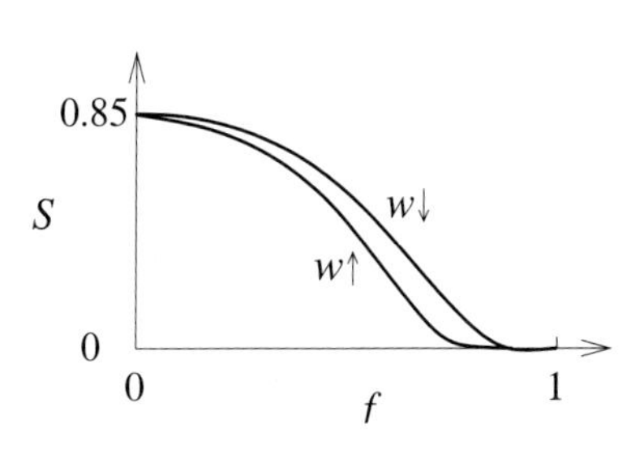

Fig. 9.4 Schematic plot of the variation of giant component size S with a fraction of removed links f. The curves labelled by $w\uparrow$ and $w\downarrow$ were obtained by the successive removal of links based on their weights, starting from low and high weights, respectively. Onnela *et al.* observed dependences of this sort in the phone call weighted network [144, 145]. Note that $S(0) < 1$ since even in the undamaged network, a fraction of nodes are in small clusters.

Finally, we approach the problem of the distribution of information flows over the network. Like Granovetter, we are interested in the information flow coming to a node from remote sources in the network. Through which sort of ties—strong or weak—do larger 'outsider information' flows pass? Fortunately, it is possible to answer the question by only analysing the structure of this weighted network. Let us accept a minimal model for the information transfer in a network. Namely, assume that each node produces the flow of news at equal rate, and that this information is sent to all other nodes through the shortest paths. Then the information flow through a link is proportional to the number of shortest paths between all pairs of nodes, which pass through this link.[3] In the phone call network, this number turned out to be, on average, high for weak ties and low for strong ties.[4] So it is the weak ties that ensure access to remote information in these networks.

[3] Since the number of remote nodes is large, one can neglect the contribution from paths between close nodes.

[4] Instead of the numbers of shortest paths, Onnela and coauthors studied the betweenness centralities of links. These quantities are closely related—the larger the number of shortest paths, the larger the betweenness centrality.

9.2 World-wide airport network

Ramified transportation systems are most naturally treated in terms of weighted networks. The weight of a link in these networks is normally defined as traffic through this link. Here we consider a very representative network of airports, where nodes are airports and links are direct flight connections. In sharp contrast to basically two-dimensional railway networks, this network is a small world. Alain Barrat and coauthors analysed the world-wide airport network including 3880 major airports and 18 810 direct flight connections [19]. This sparse network is actually small (less than 10^4), but this size was sufficient for a conclusive analysis. The weights of links were the numbers of available seats on given connections for 2002. Figure 9.5 shows a small piece of this network. In terms of its degree distribution, this network is scale-free. The exponent γ of the degree distribution is close to 2. Furthermore, the distribution of link weights was also found to be heavy-tailed, although noticeably different from a power law. In addition to the degrees, nodes in the weighted networks are characterized by their *strengths*. The strength s of a node is the total weight of its connections: $s_i = \sum_j w_{ij}$. Similarly to weights, the strength distribution in the world-wide airport network is heavy-tailed, but the network size is not sufficient to determine whether this distribution is scale-free or not.

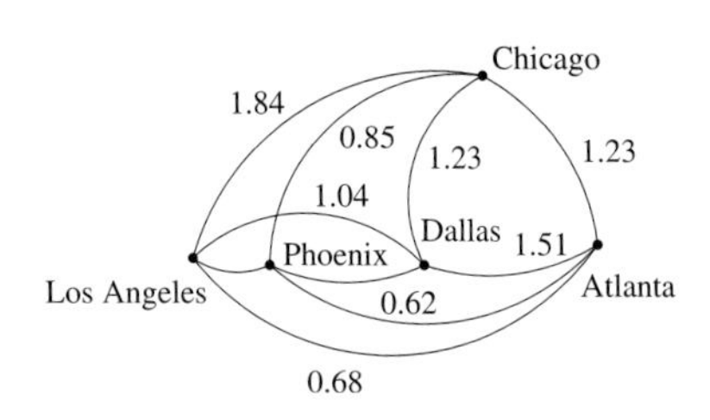

Fig. 9.5 A small part of the world-wide airport network analysed by Barrat and coauthors [19]. The weights of links are the numbers of available seats (million/year) on these direct connections.

As is clear from the definition, any weighted network can be treated as unweighted if we ignore the weights of its links. Then the first questions to be asked are: what is the relation between a weighted network and the underlining unweighted one? How do their statistical characteristics correlate with each other? In particular, how do the degrees of nodes and their strengths relate to each other? This question being especially relevant for scale-free networks, where node degrees and strengths fluctuate strongly. Barrat and coauthors measured the average strength of a node of degree q, $\overline{s}(q)$, in a wide range of degrees in several real weighted networks. They found that this dependence is close to a power law:

$$\overline{s}(q) \propto q^{\theta} \,. \tag{9.3}$$

Exponent $\theta \approx 1.5$ for the world-wide airport network. In other real-world weighted networks, exponent θ takes different values, typically, in the range from 1 to 2. That is, the hubs—nodes with many connections—usually have a high strength. In some other scale-free networks the dependence $\overline{s}(q)$ is proportional.[5] This is the case, for example, for the one-partite weighted networks of scientific collaborations [19]. In these networks, link weights indicate the intensity of cooperation within the pairs of researchers. In the weighted networks of coauthorships, the weight of a link between two researchers is proportional to the number of their joint papers. Thus a power-law $\overline{s}(q)$ curve is typical in real-world scale-free networks. What is the origin of this dependence?

[5] In this case, one can assume that $\overline{s}(q) = \langle w \rangle q$, where $\langle w \rangle$ is the average weight of a link in the network.

9.3 Modelling weighted networks

'To explain the nature of some observed phenomenon' actually means to find a reasonably realistic model demonstrating this effect. In our case, this model must provide a network in which all of the three distributions—the degree distribution, the weight distribution, and the strength distribution—have a power-law form. In addition, $\overline{s}(q)$ must be a power law or, at least, proportional to q. Barrat and coauthors proposed the first model of this kind for a growing network, exploiting a self-organization mechanism [18].

The rules for the evolution of this network are as follows.

(i) At each time step, add a node to a network. By a link of some weight, say $w_0 = 1$, attach this node to a preferentially chosen node, say node i. The probability to select a node for the attachment is taken to be proportional to the strength of this node.

(ii) Increase the strength of the node i by a constant value δ:

$$s_i \to s_i + \delta\,.$$

Do this by distributing this addition among the q_i links of this node proportionally to their weights. So the weights of these links increase:

$$w_{ij} \to w_{ij}(1 + \delta/s_i)\,.$$

This simple model is a good starting point. The reader can see that if $\delta = 0$, then the network is actually unweighted, and the model is reduced to the Barabási–Albert one. The non-zero parameter δ produces a deviation of the strength of a node from its degree and makes the evolution of weights nontrivial. For transportation, in particular, for airport networks, evolution rules (i) and (ii) are quite reasonable. Indeed, (i) a new flight connection is preferentially to a hub with higher traffic; (ii) the additional traffic due to this new connection certainly increases traffic in other flight connections from this hub. The resulting growing network has scale-free distributions of degree, weight, and strength with the same value of distribution exponent. This exponent, γ, is determined by the only parameter of the problem, that is δ.[6] As for the curve $\overline{s}(q)$ for this model network, the dependence was found to be proportional.[7] Thus this model is too simplistic to explain the power law $\overline{s}(q) \propto q^{1.5}$ observed in the real world-wide airport network.

Ginestra Bianconi found how to reproduce the law $\overline{s}(q) \propto q^{\theta}$ with exponent θ greater than 1 [27]. She proposed combining two different preferential attachment processes: (i) attachment to a node selected with probability proportional to its degree, exactly as in the Barabási–Albert model and (ii) practically the same strength–preferential attachment, as in the model described above, followed by the redistribution of weights. These two different kinds of attachment occur with complementary probabilities, p and $1 - p$. For these evolution rules, in some range of the parameters p and δ, the resulting exponent θ turns out to be greater than 1. On the other hand, there is still a wide region of p and δ in which θ is exactly 1. Thus the two classes of scale-free weighted networks can be indicated: with exponent $\theta > 1$, like the world-wide airport network, and with $\theta = 1$, like the weighted network of scientific collaborations.

For the sake of brevity, we have touched upon only local quantities here—weight and strength—and missed other characteristics. Remarkably, some characteristics of unweighted networks can easily be generalized to the case of weighted networks, and others in principle cannot. For example, in addition to the usual path length, consider the sum of the weights of the links in a path—the weight of the path. Then the notion of *an optimal path* between two nodes is naturally introduced. This

[6] The expression for the exponent is

$$\gamma = 3 - \frac{2\delta}{1 + 2\delta}\,.$$

So increasing δ results in more skewed distributions of degree, weight, and strength than for the Barabási–Albert model.

[7] Note that the model may be even simplified. Use, for example, the following rules: at each time step, increase the weight of a preferentially chosen link by some constant number and attach a new node to one of the ends of this link. This leads to the same results.

is the path with the optimal weight (minimal or maximal, depending on the problem under consideration). Optimal paths in real-world weighted networks and their models have been extensively studied. On the other hand, there is still no satisfactory analogy of the clustering coefficient in weighted networks. Note that the particular interest in weighted networks emerged rather recently, and far fewer have been studied than unweighted ones.

Motifs, cliques, communities

10

In this lecture we discuss subgraphs in complex networks. First we consider 'building blocks' of networks—*motifs*—subgraphs which are present as many copies in a network. We have already considered a simple motif—the triangle—but what is the place and role of other motifs? We also discuss ways to detect relatively weakly interconnected modules in a network—*communities*.

10.1 Cliques in networks

The k-clique is a fully connected subgraph of k nodes.[1] In particular, the triangles, which we discussed in the context of clustering, are 3-cliques. Recall that in sparse infinite classical random graphs, the number of triangles is finite, $\mathcal{N}_3 = \langle q \rangle^3/6$, where $\langle q \rangle$ is the mean degree of a node. The higher k-cliques must occur even more rarely than triangles, and therefore we can expect that the k-cliques with k greater than 3 are almost surely absent in these networks. Mathematicians have proved that this is indeed the case for sparse classical random graphs, and for them, the maximal observable clique size is 3. When are the higher cliques significant? The obvious answer is: when a network is highly clustered. This, in particular, takes place in dense classical random graphs. For us, however, widespread sparse networks are far more interesting. We have explained that sparse uncorrelated networks may have high clustering if their degree distributions decrease sufficiently slowly. Keeping this in mind, Ginestra Bianconi and Matteo Marsili inspected sparse uncorrelated networks with a divergent second moment $\langle q^2 \rangle$ and indeed found numerous (k>3)-cliques [33].

[1] For simplicity, in this lecture we consider only undirected networks.

Given that at least in some networks, k-cliques are in abundance, one can ask: how are they connected with each other? Or, what is the organization of a network in terms of its k-cliques? Hungarian researchers Derényi, Palla, and Vicsek introduced the following definition: two k-cliques are adjacent (or, one may say, 'linked' or 'connected') if they share $k-1$ nodes [69]. Figure 10.1 explains this definition in the case of 3-cliques, that is, triangles. One can replace a network by the complete set of its k-cliques connected in accordance with this definition. Derényi and coauthors focused on the architecture of this k-clique network obtained from a dense classical random graph.

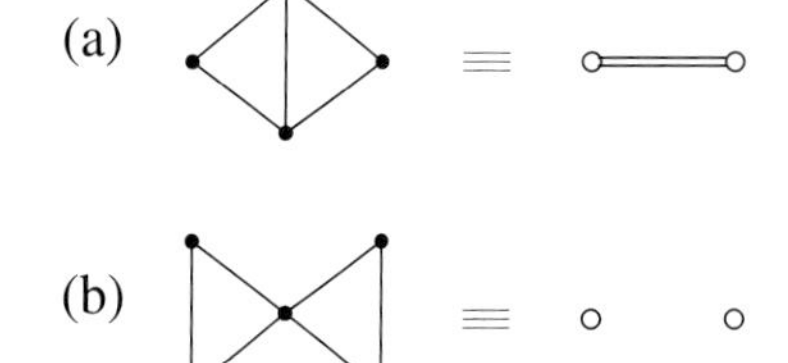

Fig. 10.1 These two 3-cliques (a) are adjacent and these two (b) are not. The open nodes and double link denote 3-cliques and a connection between them, respectively.

Let p be the probability that two nodes in the original graph are linked.[2] Assume that pN (the mean degree of a node) diverges with N, so that the the original network is sufficiently dense, and, of course, practically all of its N nodes are in a giant connected component. It turns out that the total number of k-cliques in this graph is approximately $N^k p^{k(k-1)/2}/k!$, which is a very large number in comparison with N. The degree distribution of the resulting k-clique network is Poissonian, as in the original graph. The mean degree of a node in the k-clique network, $\langle q \rangle_k$, was found to be approximately Nkp^{k-1}, which is much less than the number of nodes in this net as N tends to infinity. So the derivative network is sparse, in contrast to the original one. Derényi and coauthors showed that this sparse derivative network has a giant connected component if the probability p exceeds some critical value $p_c(k)$,

$$p_c(k)N \sim N^{(k-2)/(k-1)}\,, \tag{10.1}$$

as N approaches infinity.[3] The emergence of this giant k-clique component is called *clique percolation* [69]. So we have the three scales for a mean node degree $\langle q \rangle$ in classical random graphs: (i) $\langle q \rangle \sim 1$, a giant connected component emerges in this range; (ii) $\langle q \rangle \sim N^{(k-2)/(k-1)}$, a giant k-clique connected component emerges; and (iii) $\langle q \rangle \sim N$, where the network becomes fully connected. If $k = 2$, then scales (i) and (ii) coincide. On the other hand, if k is large, scale (ii) approaches the regime of an extremely dense network.

[2] Here we use the $G_{N,p}$ version of a classical random graph.

[3] The k-clique network resembles a classical random graph, and a giant connected component emerges when $\langle q \rangle_k$ is about 1. So, $Np_c^{k-1}(k) \sim 1$, which results in relation (10.1). Note that $p_c(k)$ depends on N.

One may say that clique percolation is about the emergence of a connected system of strongly overlapping communities of a given size in a network. We should acknowledge, however, that the constraint of fixed size and overlap looks somewhat artificial. What if communities have different sizes and overlaps? We will touch upon this problem in one of the next sections.

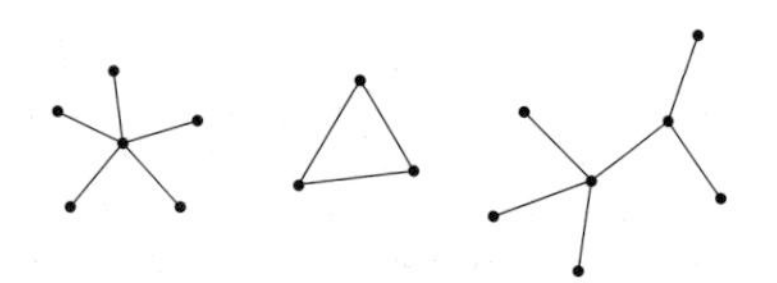

Fig. 10.2 Various stars, a triangle, and the pairs of neighbouring stars were used for the description of networks in the previous lectures.

10.2 Statistics of motifs

We started these lectures by defining a network as a set of nodes connected by links. Then we explained that the properties of a network are to a large extent determined by its local features, namely, by its degree distribution. This can be treated as the distribution of q-pointed stars in a network. We approached an even better understanding of a network by studying the statistics of triangles (3-loops) and correlations between degrees of nearest neighbours—actually, the distribution of the neighbouring star pairs. These were the simplest patterns which we used extensively in describing the network structure, see Fig. 10.2. For example, in terms of stars and triangles, the graph shown in Fig. 10.3 is described as follows: the graph contains three 1-point stars, two 2-point stars, one 5-point star, and one triangle, and so on. In essence, this was our standard approach to networks in the previous lectures. Figure 10.4 explains another way of characterizing a network: index and enumerate network motifs starting from the smallest and simplest. In particular, in

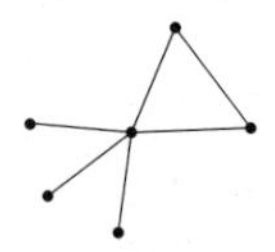

Fig. 10.3 A simple graph, which has three 1-point stars, two 2-point stars, one 5-point star, and one triangle. To present this list is actually the same as to give the degree distribution and clustering.

this figure, we show the 3-node motifs (connected 3-node subgraphs). In its ultimate form, this description implies the statistics of all isomorphic subgraphs in a given network, starting from the smallest and simplest subgraphs.[4] Usually, however, only rather small motifs are discussed, that are present in many copies in a network, which are its elementary building blocks. There are only two different 3-node motifs in undirected networks, see Fig. 10.4. In directed networks, their number is markedly higher. If reciprocal links are allowed, as in Fig. 4.5 for the WWW, then this number is thirteen [126]. Figure 10.5 demonstrates some of these thirteen motifs. Draw the rest of the 3-node motifs. In their renowned article, Milo and coauthors focused on these specific motifs in a number of real networks, including, in particular, the *E. coli* transcription network, a domain in the WWW, electronic sequential logic circuits, and food webs [126]. In addition, for each of these networks they constructed a simplified model, namely the uniformly random network of equal size, having the same sequence of in- and out-degrees. In Lecture 8 we described a randomization procedure allowing one to construct a network model of this kind, see Fig. 8.4. Milo and coauthors compared the occurrence frequencies of the motifs in all of these directed networks. It turned out that the occurrence frequencies of a given motif are very different in different directed networks and their randomized models. Moreover, the observed number of motif copies in a real network typically exceeds that for its randomized counterpart.

[4] See Lecture 1 for the definition of graph isomorphism. Note that here only 'connected subgraphs' are considered, that is, motifs cannot have separate parts.

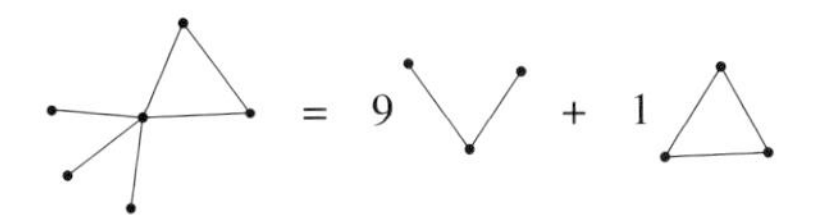

Fig. 10.4 The same graph as in Fig. 10.3 can be described as a combination of two distinct 3-node motifs. This graph also contains six 2-node motifs (two nodes and a link between them). Find the number of 4-node motifs in this graph.

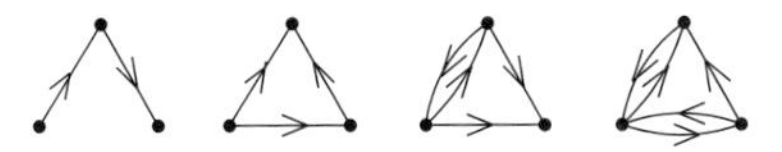

Fig. 10.5 A few of the thirteen 3-node motifs for directed networks, similar to that shown in Fig. 4.5. For the full list, see [126].

In principle, this difference is not so surprising and was expected. We have already described a strong difference in clustering in various undirected networks. So it is reasonable to characterize networks using a set of occurrence frequencies for motifs. One tempting idea stimulated the numerous studies of motifs in networks. The hope was that each specific motif is responsible for some function of a network.[5] If this be true, then the statistics of motifs would essentially describe and even determine the function of a network. This would enable researchers to predict details of the function of a network based on the distribution of motifs. Unfortunately, this program was never realized. Moreover, we believe that the idea is basically flawed. Here we present only one of the counter-arguments. Consider, for example, a protein interaction network, in which nodes are different proteins and links are pair-wise interactions between them. A motif in this network is a specific pattern of interactions between proteins. Distinct copies of the same motif contain different proteins. So, the different copies may be responsible for quite different functions in a network. Then, it is indeed impossible to strictly relate motifs to specific network functions.

[5] The reader will find a strict formulation of this concept on the home page of Uri Alon (http://www.weizmann.ac.il/mcb/UriAlon) who is one of the authors of [126]. In application to networks of transcriptional interactions, he stressed: 'We find that much of the network is composed of repeated appearances of three highly significant motifs. Each network motif has a specific function in determining gene expression, such as generating temporal expression programs and governing the responses to fluctuating external signals. The motifs also allow an easily interpretable view of the entire known transcriptional network of the organism.'

The number of different n-node motifs rapidly grows with n. It is easy to check that there are already hundreds of different 6-node motifs each of which may be present in a network in many copies. These numbers provide sufficient statistics for meaningful analysis. Using this statistics, Baskerville, Grassberger, and Paczuski investigated the frequency of occurrence of different motifs of a fixed size in a network [21]. They measured the numbers of motif copies of sizes in the range from 8 to

10 in a set of real-world networks (in the *E. coli* and yeast transcriptional regulatory networks and others). For each size, they ranked all motifs according to their amount in the network and plotted the number of copies of a motif versus the rank of this motif.[6] As is very usual in network research, the implicit hope was that this distribution would be heavy-tailed or even scale-free. Interestingly, the first publication of Baskerville and Paczuski on the statistics of motifs indicated that this was the case, but finally a rapidly decreasing distribution was reliably reported for all of the studied networks. Nonetheless, no single motif was found to be prevalent in any of these networks. Instead, sets of different motifs are particularly abundant, say, dozens or hundreds of motifs, depending on motif size n. Due to overlapping, the number of motif copies may be astronomic even in relatively small and sparse networks. For example, in the *E. coli* network investigated in that work, there were only 230 nodes and 695 links. Nonetheless, the most frequent of the 6-node motifs was present in about 10^8 copies.

[6] Researchers often represent statistical data in this way. This representation is sometimes called *Zipf's plot.* Based on the ranking, Zipf's plot provides a monotonously decreasing dependence with few noticeable fluctuations. This weakness of fluctuations allows better fitting and analysis of the data. Very similarly, instead of a degree distribution $P(q)$, empirical researchers usually plot a so-called *cumulative distribution* $P_{\text{cum}}(q) = \sum_{q' \geq q} P(q)$, which has fewer visible fluctuations than $P(q)$. You can equivalently do the following. In the spirit of Zipf, rank the nodes of a network according to their degrees and plot the rank r of a node versus its degree (or, if you want, degree versus rank). Clearly, $r(q)/N = P_{\text{cum}}(q)$, and so we have full equivalence.

One can go even further and fix not only size n of a motif but also the number of its links m [176]. Remarkably, the number of the n, m-subgraphs in a network is determined by its degree-dependent clustering $\overline{C}(q)$ and degree distribution, and so this number can easily be estimated. Even without calculations, however, it is clear that in a sparse network, densely connected motifs of a given size are less frequent than motifs with a small number of links. In that sense, cliques are the least frequent motifs in networks.

10.3 Modularity

We have demonstrated that nodes in a typical social network fall into a set of well-distinguished dense modules, see Fig. 9.1. In contrast to network motifs, these modules have no joint nodes: overlapping is absent by definition. Importantly, there are many links within modules and relatively few links between nodes in different modules. Note that, in principle, a given network may contain modules of very different sizes. Various situations are possible. In one network, modules of, say, 3 and 300 nodes can be found, and another network is of only two large modules. Consequently a modular organization is usually about a large-scale clustering of nodes in a network. A pronounced modular structure, in other words, high modularity, is typical not only for social networks but also for many others. So, how can we measure the modularity of a network? This important question is addressed frequently in network science, and it will be instructive to discuss this problem in detail.

Note that, in principle, details of the modular organization of a given network are not known *ab initio*. This modular structure—a set of modules in which nodes naturally fall—should be found. This difficult task can be carried out in two steps. Firstly, assume some particular division of the network into modules and define a measure of the 'quality' of this division. We will explain below what is a high-quality partition. It is this quality measure—a characteristic number—that is called *the modularity* of a given partition of a network. The second, most difficult step is to compare all possible partitions of the network and find the best of them, with optimal modularity. The optimal partition gives the actual set of modules in the network. The modularity for this partition is the modularity of a network. Unfortunately, the second step—optimization—cannot be done explicitly. The point is that the number of possible divisions grows rapidly as a factorial of the network size. This number is huge even in small networks. This makes the optimization problem computationally very hard, requiring some approximations and efficient numerical algorithms. The main content of hundreds of published papers on network communities consists of a frantic search for the best approximations and algorithms of this kind.

Let us focus on the modularity of a particular partition of a network, leaving the discussion of step 2 for the next section. Our immediate task is to define modularity Q for a given division of a network into a set of modules. Assume that a network of L links is divided into n modules, $r = 1, 2, \ldots, n$, with l_r links in module r. The definition should preferably provide the maximum modularity if all modules are separated from each other, and the minimum, say $Q = 0$, in a homogeneous situation, in which modules are imperceptible. Mark Newman and Michelle Girvan found a way to meet these requirements [139]. They proposed to base the definition on the comparison of two networks: a given network and its zero modularity counterpart—the uniformly random network with the same sequence of degrees as the original one. A quantity for comparison was the ratio of the total number of links within modules and the number L of links in the network. In other words, this is the probability that a link is within one of the modules. This gives the definition:

$$Q = \frac{1}{L}\sum_{r=1}^{n} l_r|_{\text{given net}} - \frac{1}{L}\sum_{r=1}^{n} l_r|_{\text{uniform counterpart}}\,. \tag{10.2}$$

In the uniformly random counterpart, the ratio $\sum_r l_r/L$ can easily be calculated since we know the degrees of all nodes. It is actually enough to know the total degrees q_r of the nodes within individual modules, where $r = 1, 2, \ldots, n$ and $\sum_r q_r = 2L$. We must find the probability $p_r = (l_r/L)|_{\text{uniform}}$ that a randomly chosen link in the uniformly random network is in module r. Similarly to the configuration model, the probability that a given end of a link is in module r is proportional to q_r. Consequently, the probability p_r is proportional to q_r^2.[7] Then $p_r = (q_r/2L)^2$. Indeed, if, for example, the entire network consists of a single module, we have the probability $(2L/2L)^2 = 1$. Then the definition of modularity takes the form:

[7] Note that in the uniformly random counterpart, multiple and self-connections are allowed.

$$Q = \sum_{r=1}^{n}\left[\frac{l_r}{L} - \left(\frac{q_r}{2L}\right)^2\right]. \tag{10.3}$$

Clearly Q is smaller than 1. If a network has strong community structure, that is, well-distinguished modules, then for this partition we have

Q close to 1. Another division of the same network gives a smaller Q. In a uniformly random network, $Q = 0$. However, this definition allows even negative Q in some situations. In particular, this can occur if multiple and self-connections are absent in a given network. For example, let a network without 1-loops be divided into N modules, each of a single node. These 'modules' clearly have no internal links, and the first term in the definition equals zero. Contrastingly, the uniformly random counterpart contains self-connections, which guarantees a positive second term. Together, these terms give negative modularity.

Instead of expression (10.3), researchers often use an equivalent one, in terms of the adjacency matrix a_{ij} and the node degrees q_i,

$$Q = \frac{1}{2L} \sum_{r=1}^{n} \sum_{i,j \in r} \left(a_{ij} - \frac{q_i q_j}{2L} \right). \tag{10.4}$$

Here the sum $\sum_{i,j \in r}$ is over all pairs of nodes in module r, and the difference $a_{ij} - q_i q_j / 2L$ is called a *modularity matrix*.[8]

Typically, positive modularity is observed in real-world networks, in the range about 0.3—0.8.[9] For demonstration purposes, here we compile only a short list of modularity values, which Mark Newman obtained for a few diverse networks of different sizes and architectures [135].

- For the karate club network of Zachary of 34 nodes, Q is 0.419.[10]
- For a metabolic network for the nematode *C. elegans* of 453 nodes, Q is 0.435.
- For a coauthorship network of scientists working on condensed matter physics of 27 519 nodes, Q is 0.723.

Importantly, it is hard to find a real-world network with low modularity. Even this short list shows that modular network architectures are widespread.

The useful modularity definition (10.2) has been implemented in hundreds if not thousands of studies. Nonetheless, there is some controversy. First of all, one may wonder: is it in principle possible to describe so complicated a feature as modularity by a single number, Q? The second point is: in the definition, an uncorrelated network with multi- and self-connections is set as a zero modularity counterpart of a given network. This choice is not that obvious if, say, the original network has no multiple connections. We noted that, strangely, when applying to networks without multi- and self-connections, the original definition may produce a negative modularity. Furthermore, for a linear chain, the definition counterintuitively gives non-zero modularity, despite the fact that chains can hardly be called modular objects.[11] In short, be careful when using this definition of modularity.

[8] To understand this form of the definition, recall that $\sum_{i,j} a_{ij} = 2L$ (see Lecture 1.

[9] Here modularity is for the optimal partition of a network into modules.

[10] The small Zachary karate club graph [184] was a reference network in numerous studies of modularity and communities. This is a one-partite network of friendships between members of a karate club at one of the US universities. A social conflict in this group divided the network into two overlapping communities, which made community indexing non-trivial. Researchers often test new community detection algorithms on this network.

[11] There is a natural way to improve this definition. Namely, impose the structural constraints of an original network on its zero modularity counterpart. For example, let an original network have no multiple connections. Then its zero modularity counterpart must also have no multiple connections. This way, however, was still not realized.

10.4 Detecting communities

There is no unique definition of a network community. In many works, the communities are natural, non-overlapping modules of networks, which we have discussed in preceding sections. Other studies suppose that communities can overlap. In the third type of study, the communities are hierarchically embedded one into another—the nested structure of communities. Finding/indexing communities is a remarkably attractive research direction in network science. How can we explain such a keen interest? The main aims of these studies are to find natural groups of nodes in a network and connections between them, to uncover and understand the coarse structure of a network, and to indicate particularly 'influential' communities. Of course, we can rather easily achieve these aims if a network is small, say, of 10, or 20, or 30 nodes. Analyse this small network visually and indicate its dense, closely connected parts—communities. Certainly, this can be done only if these communities are sufficiently distinct from each other. If, however, a network is large, say of hundreds nodes or bigger, the 'manual' detection of communities is impossible and an automated-indexing solution is the only option. For large networks, solution is an extremely time-consuming task. In addition, communities are poorly defined and often hardly distinguishable, so it is very difficult to develop a universally efficient numerical algorithm for the problem [94].

For a brief survey of various techniques used for indexing communities, the reader should address article [94]. Here we outline a few key ideas. Let us start with the problem of detecting non-overlapping communities.[12] The distinguishing feature of these communities is a relatively high density of links. The idea is to trace the motion of a random walker on the network over a long period of time. A random walker spends more time in areas with a higher density of links, and so efficiently indicates communities. One can consider different random walk processes. For example, in one process, at each time step, a walker moves from a node to any of its q nearest neighbours with equal probability $1/q$. In another process, a walker moves from a node to any of its nearest neighbours with a probability p and remains at the node with the complementary probability $1 - pq$.[13] Note that these two random walk processes differ sharply from each other. We will explain the difference in detail in one of the following lectures. Here we stress that results (a particular picture of communities) obtained using these algorithms, depend strongly on an implemented random walk process. So the idea is to attain the clearest recognition of communities in a wide set of networks by choosing some particular random walk process. Often, this is possible.

Instead of tracing a random walk, researchers usually inspect the action of a mathematical operator, which generates this random walk. Technically, this is more convenient. In mathematical terms, the problem is about the eigenvalues and eigenvectors of this operator. Moreover, the general idea of analysing the spectral properties of some specific mathematical operator defined on a network leads this field [135].

These 'spectral' algorithms allow us to extract non-overlapping communities. In their renowned paper, Mark Newman and Michelle Girvan considered another, 'hierarchical' picture of network communities [139].

[12] In the preceding section, we explained that the maximum modularity Q corresponds with the actual set of network communities. In large networks, it is in principle impossible to explicitly optimize Q over all possible network partitions. So all of the numerous algorithms for network community discovery perform this optimization approximately or, alternatively, bypass the optimization problem completely by exploiting some heuristic arguments. For example, use the fact that the density of connections is higher within communities.

[13] Here $pq_{\max} \leq 1$, where $q_{\max}$ is the maximum degree in a network.

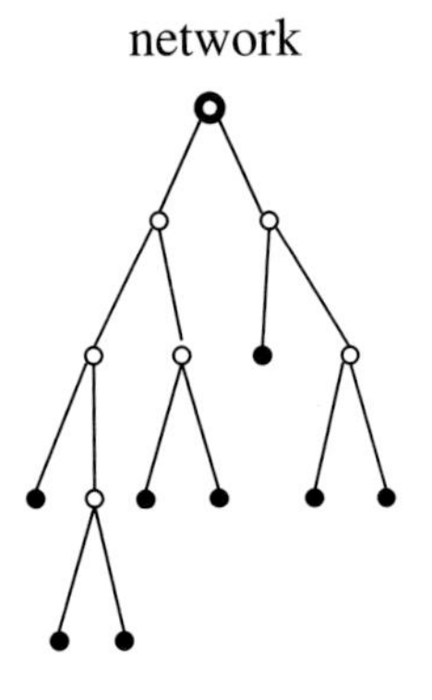

Fig. 10.6 The hierarchy of communities in a network according to Newman and Girvan [139]. The nodes are the communities, the root is the network itself, and the dead ends (filled dots) are the nodes of the network, that is, the smallest communities. Each link connects a community with a directly embedded community. In this example, the network is of 8 nodes. The two largest communities are of 5 and 3 nodes.

In this picture, communities are successively nested one into another, so that a network consists of two complementary communities, each of these communities consists of two communities, and so on. This splitting is always possible if a community has two or more nodes. This hierarchy of network communities has athe structure of a tree, shown in Fig. 10.6. Each node in this tree is a community, and the root is the original network. The dead ends of this tree (excluding the root) are the smallest possible communities, which cannot be further split. Clearly, these dead ends are the nodes of a given network. So the number of dead ends in this dendrogram (excluding the root) coincides with the number of nodes in the network.

The algorithm of Newman and Girvan uses ideas which we have already discussed in the context of the 'strength of the weak ties' concept. Recall that according to Granovetter the connections between communities in a network (interlinks) carry larger information flows than links within communities (intralinks). The information flow through a link may be roughly estimated as the number of shortest paths between pairs of nodes, that pass through this link. So, the numbers of shortest paths passing through interlinks is greater than those passing through intralinks. One can believe that this is the case, not only in social networks, but generally. Newman and Girvan used this feature directly. They proposed to uncover communities by progressively removing the links that carry the highest numbers of shortest paths. Instead of these numbers, they used the link betweenness centrality, which we discussed in Lecture 5. In more strict terms, the algorithm of Newman and Girvan is as follows [139].[14]

[14] We assume that a network is one-partite and undirected.

(i) Compute betweenness centrality for all links in a network.
(ii) Remove the link with the highest betweenness centrality.
(iii) Recalculate betweenness centrality for the remaining links.
(iv) Repeat from step (ii) until no links remain.

This algorithm progressively divides a network into a set of separated clusters. Note step (iii), which is the recalculation of the betweenness centrality of all remaining links after each removal.

Why do we have to perform these time-consuming repetitive recalculations instead of consecutively removing links with the highest betweenness centralities in the original network? The authors explained the necessity of the 'recalculation step' using a simple example. Suppose that two of the communities in a network are connected by two links, one of which has the highest betweenness centrality in the network, the second having a very low betweenness centrality. Removing these two links uncovers the obvious division into these two communities. It is clear that the link with the highest betweenness must be removed immediately. After removal, recalculation will give a high betweenness centrality for the second link. So this link will also be quickly removed, and the natural division will be readily uncovered. On the other hand, without the recalculation, the second, low betweenness centrality link would be deleted much later, after many other links. This would hamper the discovery of these communities or even make it impossible.

As the algorithm runs, the number of links decreases from L, which is the total number of links in the original network, to 0, and, in parallel, the number of extracted communities increases from 1 (the network itself) to N (the total number of nodes). The result is a dendrogram presented in Fig. 10.7, where vertical lines show communities. Note that this is a more informative representation of the same network community structure than in Fig. 10.6. For each number of removed links, this dendrogram presents a natural partition of a given network (see, for example, the intersection of the dendrogram and the dashed line in Fig. 10.7). So the algorithm provides a set of less than L partitions into communities, say, n partitions. This number is far smaller than the total number of all possible partitions for a given network, which is a good point. Thanks to the smallness of this set, we can quickly compute a modularity Q for each of these n specific partitions and select the 'best' partition having the highest modularity. For this, use the formulae from the preceding section. Thus, the Newman–Girvan algorithm finds the optimal division of the network into non-overlapping communities, but also uncovers the hierarchical organization of network partitions—the hierarchy of nested communities, see Figs. 10.6 and 10.7. We stress that the Newman–Girvan algorithm is approximate. It is based on heuristic arguments. Indeed, the key assumption about the high betweenness centralities of the interlinks is only a hypothesis. At best, this property can only be valid statistically, with less than 100 per cent probability. As a result, we have the chance to miss the optimal partition. Nonetheless, despite the strong assumption behind this famous algorithm, it uncovers the community structures of various networks with reasonable success. The algorithm correctly finds the modular structure in clear situations and approximately reproduces the results of more sophisticated algorithms in more difficult cases.[15]

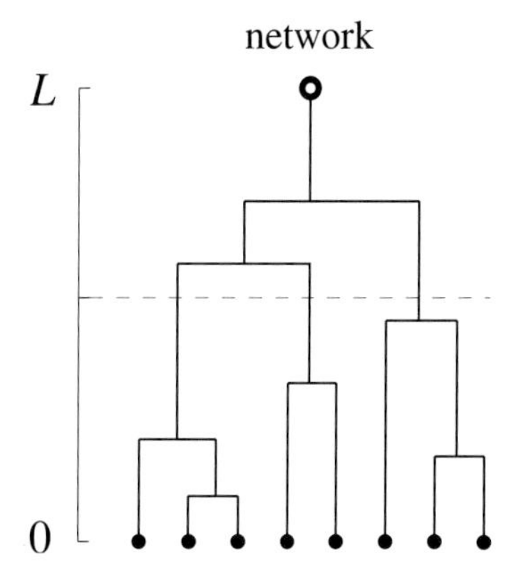

Fig. 10.7 Another representation of the community structure of the same network as in Fig. 10.6. This dendrogram is a direct result of the application of the Newman–Girvan algorithm [139]. The vertical axis shows a current number of links in the network during the execution of the algorithm. For example, at the level indicated by the dashed line, the algorithm shows three communities.

[15] There are situations, where the algorithm fails: e.g. three equal cliques interconnected by three links.

The idea that each node in a network enters into only one community, can easily be disputed. Actually, this is a very simplifying assumption. For example, in developed social networks, an individual is inevitably involved in many social, cultural, and professional activities, which results in simultaneous participation in a number of communities. This overlapping of communities severely hampers the uncovering of the community structure of a network. A few attempts have been made to fulfil this difficult task. Let us consider one of them.

In 2005, Palla, Derényi, Farkas, and Vicsek proposed defining communities using their concept of k-clique percolation [147]. We have explained this notion. Let the number k be given. According to Palla and coauthors, a community is a connected k-clique component. In other words, it is a subgraph consisting of all the k-cliques that are mutually reachable through $(k-1)$-node overlaps. It is clear that two communities of this kind can overlap, but the overlapping is not by more than $k-2$ nodes. Figure 10.8 explains the idea of Palla and coauthors. Unfortunately, community structure strongly depends on k. So the problem is

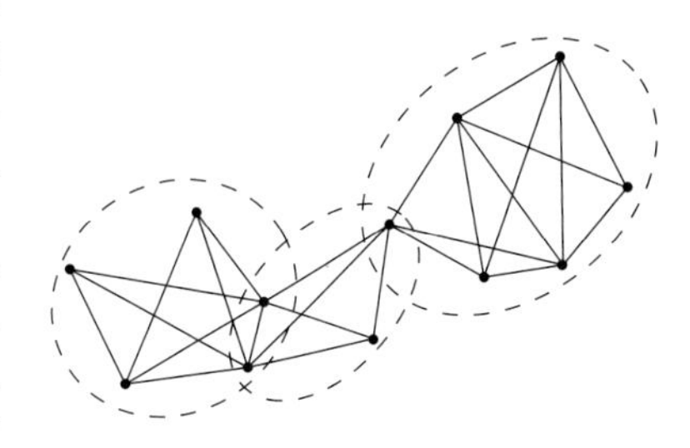

Fig. 10.8 Overlapping k-clique communities according to Palla and coauthors [147]. In this example, $k=3$. The networks consists of three communities: of two, one, and three 3-cliques.

how to choose this number. Both large and small k lead to an apparently incorrect community structure. Indeed, if k is too large, the resulting communities will be separate. On the other hand, if k is too small, the communities will be unrealistically large. If, for example, $k=2$, then these communities are simply connected components, and so a network of a single connected component is one community. Therefore, the researchers chose some intermediate value (k = 4, 5, or 6, depending on the network), which gave a sufficiently rich community structure with numerous overlaps by 1—30 nodes. They investigated this community structure in a number of highly clustered biological and social networks. Remarkably, the size distribution of these communities was found to have a power-law large-degree part, with exponent in the range 2.0—2.6. This power-law distribution was observed even in non-scale-free networks. Furthermore, one can naturally introduce the 'degree' of a community. This is the number of its overlaps with other communities, that is, the number of nearest neighbours. The observed degree distribution for the communities is also scale-free.

The work of Palla and coauthors indicates the difficulty of this rapidly progressing research field. It is not easy to arrive at a uniform community picture since different researchers can understand and define communities very differently.

10.5 Hierarchical architectures

Using tree or, equivalently, dendrogram schemes similar to those in Figs. 10.6 and 10.7 allows one to describe a hierarchy in the set of nodes in a network. In particular, in the tree in Fig. 10.6, the hierarchical position of a node is naturally characterized by the distance of this node from the root. Importantly, the forms of tree or dendrogram in these figures reflect the global organization of a network. The difference between these trees for different networks may be spectacular.[16] Figure 10.9 shows two contrasting examples of dendrograms for two different networks. Apparently, the right dendrogram indicates the presence of well-distinguished communities in the network and, one may say, its strictly 'hierarchical architecture'. In other words, this net contains a clear hierarchy of communities, unlike the network producing the left dendrogram. Hierarchical architectures of this kind have been reported in a number of real-world networks, for example, in networks of metabolic reactions [155, 154].

Deterministic graphs in Fig. 7.6 also have a clear hierarchical design. For these deterministic graphs, one can easily find degree-dependent clustering $\overline{C}(q)$. This quantity is inversely proportional to degree: $\overline{C}(q) \propto 1/q$. It is tempting to guess that a decreasing dependence $\overline{C}(q)$ is a generic feature of hierarchically organized networks. We believe that this is not correct. Many networks with degree–degree correlations show this dependence even without being hierarchical.

[16] Aside from the Newman–Girvan algorithm, there are a number of ways to build tree schemes of the node hierarchy for networks. The resulting trees and dendrograms also depend on the algorithm used.

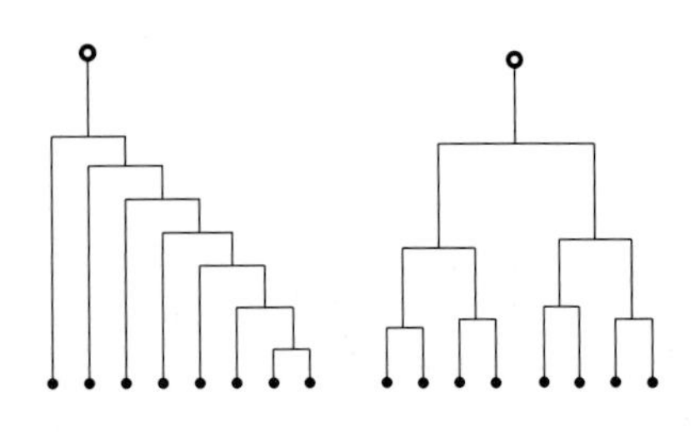

Fig. 10.9 Two contrasting examples of dendrograms for two networks. The right dendrogram indicates the hierarchical organization of a network.

Navigation and search

11

The starting point in the science of networks—Euler's solution of the Königsberg bridge problem—addressed walks on a graph. In comparison to lattices, complex networks provide remarkably rich environments for walks and related processes. In this lecture we consider the specifics of walks, navigation, and search processes in networks of various architectures and geometries.

11.1 Random walks on networks

For the sake of brevity, we touch upon only one version of a random walk. At each time step, a walker moves randomly from node to node along the links of a graph. In a so-called *simple random walk*, walker moves from a node to any of the nearest neighbours of this node with equal probability (the drunkard's walk). We can start a walk from a given node in a network and keep track of how it evolves: namely, how the probability of finding the walker at various nodes varies, how and when the walker returns to the starting node or moves off to infinity, and so on. Since a walker spends more time in dense parts of a network, we can use this process to explore the heterogeneous structure of the network. We mentioned that random walk based algorithms are applied to the detection of communities. Numerous chaotic 'one-particle' processes taking place on networks can be treated as random walks.[1] Recall, for example, Internet traffic. Optimally, packets in the Internet traffic are routed along the shortest path between a source and destination. In reality, however, the routing of packets is often so chaotic that it rather resembles random walks.

[1] Physicists use the term 'one-particle processes' for processes in which a single particle participates or particles move independently.

Suppose that a graph (or lattice) of size N is regular, that is all nodes have equal degrees. Then, at infinite time, we will find the walker at any node with equal probability, $1/N$, irrespective of the starting point of the walk. Heterogeneity changes this result. Let a network have no separate parts. Then it is easy to show that the probability of finding a walker at a node of degree q approaches the following final value:

$$p_{\text{fin}}(q) = \frac{q}{Q}, \tag{11.1}$$

where $Q = \sum_i q_i$ is the total degree of nodes in a network. In a random network, $p_{\text{fin}}(q) = q/(N\langle q \rangle)$. Figure 11.1 explains this formula. Note that this result does not depend on the architecture of the network. Relation (11.1) is correct only for undirected networks. The structure

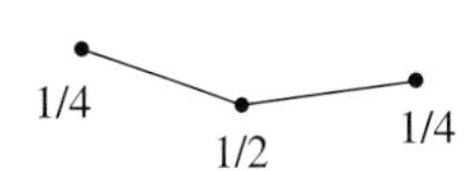

Fig. 11.1 The numbers indicate the final probabilities of finding a random walker at the nodes of this graph. In the final stationary state, for each link, the probabilities of moving in the opposite directions between the end nodes must be equal: $1 \cdot 1/4 = 1/2 \cdot 1/2$.

of connections in a directed network is more rich, as we observed for the WWW. There may be an area or areas connected to the rest of a network only by incoming links. These areas (components) are deadly traps for a random walker. If a network is finite, then the final probability p_{fin} is non-zero only in these traps. Consequently, random walks, enable us to detect specific components in directed networks.

The simple relation (11.1) leads to another important result for random walks. The problem is: how soon will a walker return to the origin of his walk? In 1947, Mark Kac derived his famous formula which directly relates the average first-return time $\langle\tau\rangle$ to the origin of a walk to the probability p_{fin} of finding a walker at this point at infinite time: $\langle\tau\rangle = 1/p_{\text{fin}}$. From the Kac formula, we readily obtain the value of the average first-return time to starting node i of degree q_i [141]:

$$\langle\tau_i\rangle = \frac{\langle q\rangle}{q_i} N \,. \tag{11.2}$$

This time diverges in an infinite network. Note, however, that for hubs, the mean return time may take a rather moderate value even for extremely large networks. Let us estimate, for example, $\langle\tau_{\text{h}}\rangle$ for a hub in a scale-free recursively growing network with a degree distribution $P(q) \sim q^{-\gamma}$. Let exponent $\gamma = 2.1$ and $N = 10^{11}$. We have shown that the largest number of connections of a node in this network is of the order of $N^{1/(\gamma-1)}$. Then for this node we have the average return time of the order of $N^{(\gamma-2)/(\gamma-1)} = 10$. Even if a network is of astronomical size, $N = 10^{33}$, we obtain a rather moderate number $\langle\tau_{\text{h}}\rangle \sim 1000$.

We can naturally generalize the average first-return time $\langle\tau_i\rangle$ to $\langle\tau_{ij}\rangle$. This is the average first-passage time of a walk which starts from node i and passes node j. This quantity characterizes separation between nodes i and j in a network from the point of view of a drunken sailor who started his walk from node i. In respect of stochastic processes of this kind, $\langle\tau_{ij}\rangle$ is a more adequate characteristic of a network than a shortest-path distance. For example, for random traffic, $\langle\tau_{ij}\rangle$ is more relevant. Note that, in heterogeneous networks, $\langle\tau_{ij}\rangle \neq \langle\tau_{ji}\rangle$ [141]. This asymmetry allows us to find which one of two nodes is more rapidly approachable by random walking.

Let us focus on infinite networks. The next natural question about a random walk is: what is the probability that a walker will return to the starting point at all? Is there a chance of escaping to infinity without returning to the origin? Apparently, the answer depends on the number of ways of escaping to infinity, that is on the dimensionality of a medium for walking. It is easier to escape in a highly-dimensional lattice than, say, in a one-dimensional chain. According to Pólya's theorem (1921), there are two possible situations for walks on a D-dimensional medium.[2]

(i) If $D \leq 2$, then a walk certainly returns to the starting point, that is the walk is *recurrent*.

(ii) If $D > 2$, a walk has a finite probability of escaping to infinity without returning to the starting point, that is, the walk is *transient*.

[2] George (György) Pólya (1887–1985) was a famous Hungarian mathematician.

Therefore random walks on small worlds (infinite-dimensional networks) are transient.[3]

Physicists traditionally study a more detailed characteristic of random walks, namely, an autocorrelation function $p_0(t)$, which is the probability of finding a random walker at the starting node after t steps. We have already mentioned that for random walks on infinite D-dimensional lattices and fractals, $p_0(t) \sim t^{-D/2}$. Consequently for small worlds, we should observe a slower decay of the autocorrelation function than any power law. In particular, for classical random graphs, $p_0(t) \sim \exp(-Ct^{1/3})$, where C depends on $\langle q \rangle$ [157].[4] Unfortunately, in contrast to lattices, this specific decay can be observed only in unrealistically large networks. Their diameter, $\ln N$, must be very much greater than 1. It is easy to show, however, that in finite networks, the decay of the autocorrelation function is exponential $p_{0i}(t) \sim e^{-t/\tau_i}$, where the relaxation time τ_i, in principle, depends on a node. Somewhat counter-intuitively, it turns out that, in contrast to the average first return time, eqn (11.2), this dependence is rather weak. Even in scale-free networks, where degrees vary by a few orders of magnitude, τ_i of all nodes were found to be surprisingly close to each other [141].

[3] This feature—transience—is closely related to another classical result for random walks. If $D > 2$, then after t steps, a random walker visits, on average, n_v different nodes, where n_v is a finite fraction of t. In particular, this is the case for small worlds.

[4] One can show that the decay of the autocorrelation function is determined by low-degree nodes, so this law is also valid for other uncorrelated networks, including scale-free. For this, a network must have nodes with less than three connections, otherwise, the decay is exponential.

11.2 Biased random walks

Let us return to the routing of packets in the Internet. Because of a number of reasons, which we will touch upon in the next lecture, the traffic is usually far from optimal. So the path of a packet from source to destination may be very far from the shortest path. In this regime, a packet moves at random. The key point is that the destination node attracts a packet, so that the random walk is biased, and the bias is towards the target. Importantly, the parameters of this walk depend strongly on the magnitude of the bias. It is interesting to understand how much bias can change the character of a random walk. The presence of bias means that the probability of a jump from a node in the direction of the destination exceeds the probability of a jump from this node in the opposite direction, see Fig. 11.2. One can show that, for biased random walks on networks, the relevant bias is exponential, that is the ratio of the probabilities is fixed:

$$\frac{p(i;\ell \to \ell-1)}{p(i;\ell \to \ell)} = \sqrt{\lambda} = \frac{p(i;\ell \to \ell)}{p(i;\ell \to \ell+1)}, \tag{11.3}$$

where $\sqrt{\lambda}$ is some number greater than 1. The square root is introduced for the sake of convenience, since $p(i;\ell \to \ell-1)/p(i;\ell \to \ell+1) = \lambda$.

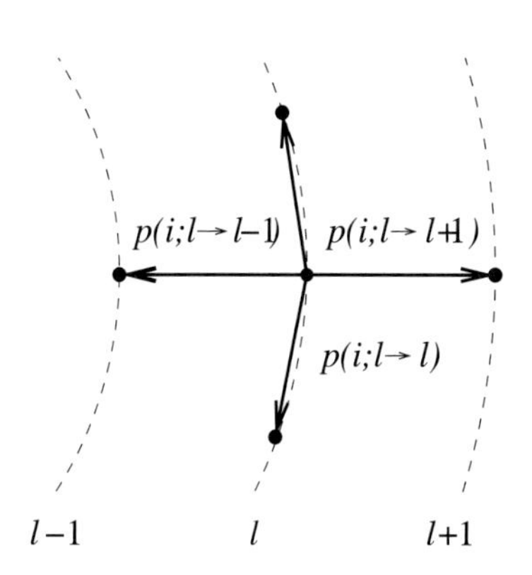

Fig. 11.2 The nodes in a network are additionally labelled according to their shortest-path distance ℓ from a destination (node-attractor). In a biased random walk, the probability of a jump from a node depends on the direction of this jump. Namely, $p(i;\ell \to \ell-1) > p(i;\ell \to \ell) > p(i;\ell \to \ell+1)$.

Here we discuss only one characteristic of the biased random walks on a network, namely, the mean return time, also averaged over all nodes, $\langle \tau \rangle$. Specifically, the dependence of $\langle \tau \rangle$ on the network size is important. It turns out that the dependence $\langle \tau \rangle(N)$ changes sharply at a certain, critical value of the bias parameter, λ_c, [167, 23]. This critical value exactly coincides with the mean branching in the network, $\lambda_c = \overline{b}$. There are three contrasting regimes:

(i) If $1 \leq \lambda < \lambda_c$, then $\langle \tau \rangle \sim N^{\ln(\overline{b}/\lambda)/\ln \overline{b}}$.

(ii) If $\lambda = \lambda_c$, then $\langle \tau \rangle \sim \ln N$.

(iii) If $\lambda > \lambda_c$, then $\langle \tau \rangle$ approaches a finite value at large N.

The regime below λ_c is delocalization. In this regime, in an infinite network, a walker escapes the bias. That is, the walks are transient. At the critical value λ_c, a *localization transition* occurs, and above this point, a walker is trapped by the bias. In this regime, the walks are recurrent, that is most walks approach the destination in a finite time. The localization transition disappears if $\overline{b} \sim \langle q^2 \rangle$ diverges, that is, hubs hamper localization. One should stress that routing in the Internet is much more complicated than this idealistic biased random walk model, which prevents meaningful comparison. In reality, packets travel not quite independently of each other, queueing at network routers.

11.3 Kleinberg's problem

Recall Milgram's small-world experiment, in which he found the famous six degrees of separation between individuals. We described the experiment in detail in Lecture 1. Note that these six degrees are not the length of the shortest path between two persons in the network of acquaintances but rather a rough estimate from above. Indeed, what was the essence of Milgram's idea if we ignore less important details? The participants in the experiment were asked to forward a letter to those of their acquaintances who were closer to the target person. The target's address was known, and the 'closer', in this idealization, simply means 'geographically closer'. Each participant knew the addresses of his or her acquaintances, and so it was easy to select a proper recipient. In this search process (searching for the shortest route to the target), all participants used very reduced, local information, namely the coordinates of their acquaintances, and, of course, the target's address. The participants had no idea about the full structure of their network.

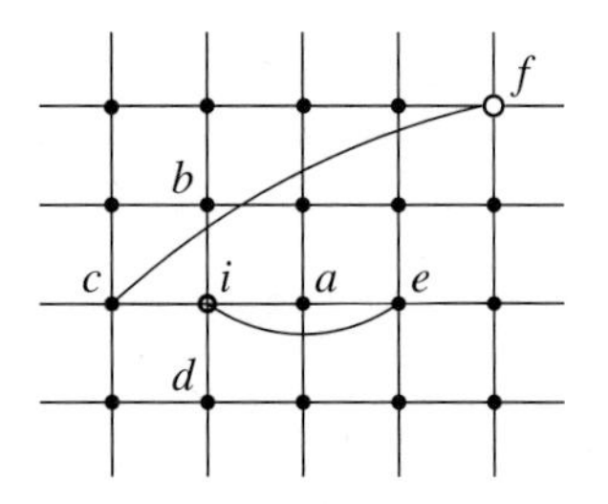

Fig. 11.3 How the greedy algorithm works. Let the goal be to reach the target (node f), starting from node i, in the shortest time. The greedy algorithm results in a four-step path passing through node a, while the shortest path through node c is of only two steps.

Remarkably, Milgram's algorithm is the standard one in computer science. Computer scientists call algorithms of this kind *decentralized search algorithms*. Specifically, this is the simplest process—*the greedy search algorithm*: repeatedly pass to the neighbour node, geographically closest to the target. Importantly, the nodes use only local information. We can roughly model the network of acquaintances in Milgram's experiment by a two-dimensional lattice with shortcuts—a small-world network. Figure 11.3 explains the greedy algorithm for this network. Each step is to the nearest neighbour, geographically closest to the target. So, with each step, we are getting geographically closer to the destination. This restriction—strong bias towards the target—simplifies the algorithm and makes it ultimately efficient and quick. On the other hand, 'geographically closer' does not mean 'closer in the network'. So moving in this manner, we easily miss the shortest path to the target even in very simple situations like that in the figure. We conclude that

Milgram's six degrees of separation are simply the delivery time of the greedy algorithm for his specific problem.

How quickly can we find a target in a network by using the greedy algorithm (greedy routing)? Or, equivalently, how easily can we navigate through a network? The answer depends strongly on the architecture of the network and on its size. A relevant characteristic quantity for a network is an average delivery time of the algorithm, $\overline{\tau}$. Here the average is over all pairs of nodes–starters and nodes–targets in the network. Let the substrate of the network be a D-dimensional lattice of $N = L \times L \times \ldots \times L$ nodes. Then $\overline{\tau}$ depends on L and D. The size dependence $\overline{\tau}(L)$ is the main issue of interest. For a lattice without shortcuts, clearly, $\overline{\tau} \sim L$, and it takes lot of time to reach the target. Apparently, added shortcuts diminish delivery times, but how much?

In 2000 computer scientist and applied mathematician Jon Kleinberg considered this problem for a generalization of Watts–Strogatz small-world networks [110–112]. Originally, his network was based on a two-dimensional lattice substrate, but here we assume it to be D-dimensional. Each node of the lattice has a shortcut to a node at the Euclidean distance ℓ drawn from a power-law probability distribution, $p(\ell) \propto \ell^{-\alpha}$.[5] If exponent α equals zero, then shortcuts connect uniformly randomly chosen nodes, and so we arrive at one version of a small-world network. If α is large (short-range shortcuts), then the network, in effect, approaches a D-dimensional lattice, and $\overline{\tau} \sim L$. Long-range shortcuts surely decrease the delivery time, but how much? In particular, uniformly distributed shortcuts ($\alpha = 0$) give a small chance of getting closer to the target and so they they do not substantially improve navigation compared to a pure lattice. Kleinberg studied the size dependence $\overline{\tau}(L)$ in the entire range of exponent α values, from zero to infinity, and found dramatically different dependences. Figure 11.4 shows the resulting average delivery time versus exponent α for a network of a given size. The main finding was that the delivery time of the greedy algorithm has a deep minimum at $\alpha = D$. Kleinberg proved that at this unique point, the size-dependence of the delivery time is very slow—polylogarithmic, $\overline{\tau} \sim \ln^2 L$, while the delivery time increases as a power of L, $\overline{\tau} \sim L^x$, at all other values of α. Exponent x approaches zero as $\alpha \to D$. This sharp difference allowed Kleinberg to introduce the notion of navigability. In terms of Kleinberg, a network is *navigable* if the greedy algorithm provides rapid navigation, that is if $\overline{\tau}(L)$ increases slower than any power law of L.

Kleinberg's construction provides a navigable network at one point, $\alpha = D$. Note that even at this point, the delivery time much exceeds the average shortest-path length, $\ln^2 L \gg \ln L$, $L \gg 1$. However, for realistic network sizes, the logarithm is not large, and the difference is not impressive. So at this, and only at this, point the greedy algorithm does a good job. Although we do not show how these results were obtained, it is worthwhile explaining a few key issues. Remarkably, this system is a small world, that is infinite-dimensional, in a wider range of α than one could expect. This takes place for $0 \leq \alpha \leq 2D$. In this

[5] In principle, the number of shortcuts per node can be taken to be equal to an arbitrary finite number, without loss of generality.

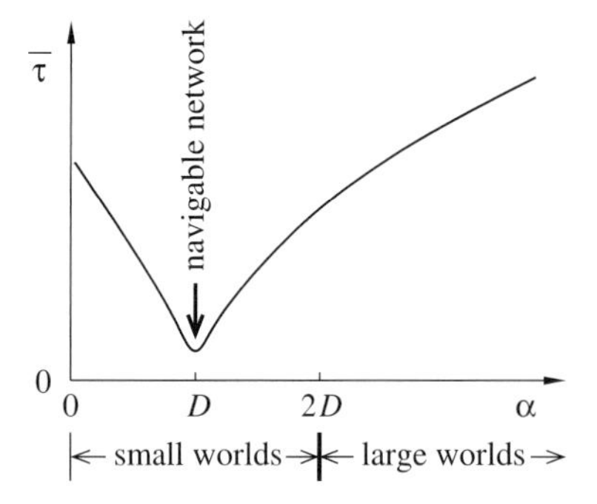

Fig. 11.4 Schematic plot of the average delivery time provided by the greedy algorithm versus exponent α according to Kleinberg [110]. The size of the network is fixed. When the dimensionality D of the underlying lattice is below or equal to $2D$, the network is a small world.

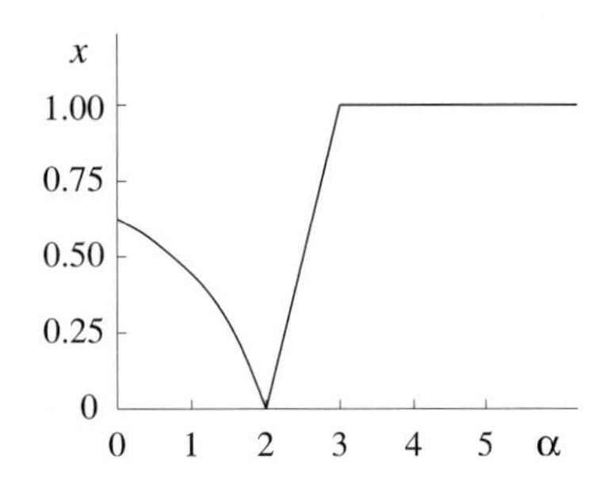

Fig. 11.5 Delivery time exponent x ($\overline{\tau} \sim L^x$) versus exponent α for the two-dimensional Kleinberg model according to Carmi *et al.* [51] and Cartozo and De Los Rios [52]. The network is navigable at $\alpha = 2$.

region, except when $\alpha = D$, $\overline{\tau}$ is dramatically greater than the average shortest-path length. Compare the analytical expressions for asymptotic dependencies $\overline{\tau}(L)$ at various α [51, 52]:

$$\overline{\tau}(L) \sim \begin{cases} L^{(D-\alpha)/(D+1-\alpha)} & 0 \le \alpha < D\,, \\ \ln^2 L & \alpha = D\,, \\ L^{\alpha-D} & D < \alpha < D+1\,, \\ L & \alpha > D+1\,. \end{cases} \tag{11.4}$$

Figure 11.5 illustrates these formulae showing the dependence $x(\alpha)$ of the exponent in the power law $\overline{\tau} \sim L^x$.

Computer scientists widely discussed Kleinberg's major result—the optimal navigation time $\overline{\tau} \sim \ln^2 L$ for greedy routing. Importantly, they could not reduce this delivery time by using more sophisticated algorithms. Suppose that we try to 'improve' the greedy algorithm by additionally accounting for the positions of the second nearest neighbours of a node or in an even larger neighbourhood, say, involving n nodes.[6] This will certainly reduce the path to the target. Its length will be closer to that of the shortest path. However, estimating the effective time for navigation, we must multiply the length of the path by n. It turns out that this product cannot be less than of the order of $\ln^2 L$. In that respect, $\overline{\tau}$ is a basic network characteristic.

[6] Note that this 'improvement' uses non-local information about the network.

11.4 Navigability

We missed one important point in Milgram's experiment. Many chains of acquaintances in the experiment were not completed, so that a fraction of the letters failed to reach the target. That fraction was so large that at first Milgram's attempt was unsuccessful. This is not the case for networks of the kind shown in Fig. 11.3, that is for regular lattices with added shortcuts. On these networks, all search chains generated by the greedy algorithm surely reach the target. We can easily modify the model network to reproduce broken Milgram's chains. For example, remove from the underlying lattice a fraction of nodes or links. This results in a finite probability that the greedy algorithm will lead a searcher (navigator) into a trap—nodes without neighbours geographically closer to the target. The greedy algorithm does not allow a searcher to return backwards for a few steps to try to avoid the trap. When the number of removed nodes or links is large, the probability of being get stuck in this way can be high. In this situation, the greedy search algorithm is, of course, utterly impractical, and efficient searching and navigation are impossible without global knowledge. It is natural to call a network *searchable* if a sufficiently large fraction of the search chains in this net are successful [180]. In more quantitative terms, in a searchable network, the probability of an arbitrary search chain reaching its target must exceed some given value. This threshold value determines the quality of searching.

Whether the model network is searchable or not depends on its size, on how much the underlining lattice is damaged, and on the distribution of shortcuts (exponent α). Here we will not discuss the form of the area of searchability and other details. Instead, let us look at efficient navigation on networks from a more broad perspective. Instead of damaging the underlying lattice, we can use any other network substrate embedded in Euclidean space. For example, this can be a so-called random geometric graph.[7] Furthermore, instead of the somewhat artificial division of a network into a substrate and shortcuts in this construction, we can simply assume that the nodes of a given network are located in Euclidean space. The Euclidean coordinates of the nodes play the role of hidden variables allowing greedy routing. Even more generally, we can place the nodes of a network in any space where the notion of distance between points is defined [160,35], and so where it is possible to introduce the coordinates of the nodes. In mathematics these spaces are called *metric spaces*. A standard example is Euclidean space. Another example is provided by a, say, binary tree, whose leaves play the role of the points of this discrete metric space, see Fig. 11.6. The distance between two points is defined as the degree of separation of these points from their closest common ancestor in the tree. In social networks, these hidden 'tree metric spaces' naturally represent hierarchies of successively embedded communities to which an individual belongs [180].

[7] Choose N points uniformly at random in a bounded region in D-dimensional Euclidean space and connect each two points at a distance less than some fixed number. The resulting network—*a random geometric graph*—has a rapidly decreasing degree distribution and D-dimensional geometry [67].

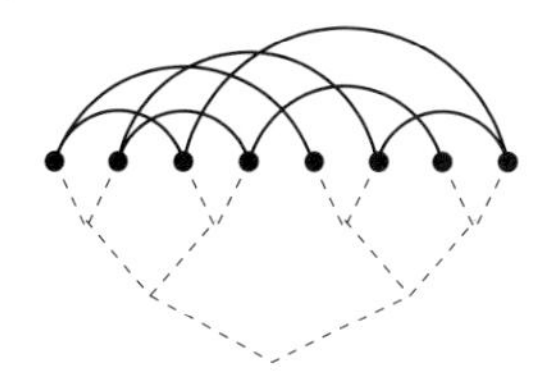

Fig. 11.6 This network (links are shown by arcs) is embedded in a binary tree metric space (dashed lines). In this discrete space, the distance between, for example, two left-most nodes equals 1, while the distance between the right-most and left-most nodes equals 3.

Marián Boguñá, Dmitri Krioukov, and kc claffy—scientists from Barcelona and San Diego—conjectured the presence of hidden metric spaces behind many real-world networks [35]. These underlining spaces are, in fact, fundamentally necessary. Individual nodes in such networks as cellular and many others, in principle, cannot have any global view of a network, and so without hidden metric spaces, nodes could not route signals, messages, and so on to intended targets. In a broad sense, it is the existence of these invisible metric substrates that makes a network navigable. In some networks, these underlining spaces are easily identified, for example, in airport networks. For other networks, these spaces are still unknown. Researchers hope to extract them based on the degree of similarity of nodes in networks, using their various intrinsic characteristics. Up to now this task has not been fulfilled.

The goal is particularly ambitious for Internet routing. Currently, routing protocols are actually based on a full knowledge of the Internet connections. Databases of individual routers—*routing tables*—store the routes to a large number of particular destinations in the network. This information is used by routers for forwarding messages by, roughly speaking, shortest paths.[8] The problem is that in the exponentially growing Internet, the total size of huge routing tables needed for this 'optimal' routing also grows exponentially with time. The urgent challenge is to find a way to cardinally reduce the routing tables to avoid rapidly approaching 'information processing overhead'. Greedy routing protocols, if realized, could resolve this serious problem.

[8] Interestingly, a typical Internet route follows a standard pattern. In the first half of the route, a message is progressively forwarded to more and more highly connected routers. In the second half, the message moves to fewer and fewer connected routers.

Efficient greedy routing assumes (i) sufficiently short routes to destinations and, necessarily, (ii) a high percentage of successful routes. Refining Kleinberg's definition, one can call networks with efficient greedy routing navigable [35]. Recall that Kleinberg's network is navigable at a single point, exponent $\alpha = D$, which makes its navigability rather particular. In contrast to this, hidden-metric-space-based models of scale-free networks turn out to be navigable in a remarkably wide range of parameters [34]. This strengthens the case for the navigability of major real-world networks.

11.5 Google PageRank

Over half a million Google servers (in 2009) in a few dozen locations around the world do terrific work. They (i) permanently crawl the WWW, fetching web pages, (ii) store many copies of its open content, (iii) index this colossal array of text data, and (iv) process an astronomical number of search queries, returning results in a ranked (that is most informative) order.[9] Here we discuss an important part of Google's technology, namely the ranking of results—*the Google PageRank* [43].

[9] See a more detailed discussion 'How Google works' on http://www.googleguide.com/google_works.html.

A search query is a set of words and numbers. The result of a query is a ranked list of the addresses of pages containing this combination. The PageRank algorithm estimates the relative importance of a web page based on its popularity in the WWW. In this way PageRank ranks all pages in the WWW. After computation, this global ranking is used for the ranking of entries in the list of results of each search query. How does the Google PageRank work? The key idea is that the popularity of a web page is proportional to the number of times a crazy web-surfer visits this page when randomly surfing the WWW. So the problem is essentially reduced to a random walk on a directed network. In one respect, this random walk differs from the one on an undirected network that we discussed earlier. In directed networks, there may be clusters connected to the remaining part of a network only by incoming connections. If a walker moves only following directed links he will be trapped finally in one of these clusters. To avoid this, we have to allow our imaginary surfer to restart the process, say, from a random node. Let r_i be the final stationary probability of finding the surfer on node i at infinite time. Then the page with the highest r receives the top rank. It is easy to show that these probabilities satisfy the following set of equations:

$$r_i = \frac{p}{N} + (1-p) \sum_{j:j \to i} \frac{r_j}{q_{\text{out},j}} . \tag{11.5}$$

Here N is the size of a network, $q_{\text{out},j}$ is the out-degree of node j, the sum is over all incoming connections of node i, and finally p is the probability that, instead of moving to one of its nearest neighbours, the walker jumps to a node chosen uniformly at random. The probability $p \neq 0$, which is the only parameter of this algorithm, was claimed to be about 0.15. It is clear why only out-degrees are present in the sum on the right-hand side of this equation. Indeed, a walker escapes from a node along each of the outgoing links with equal probability.

Equations (11.5) are solved numerically by iteration. This is a simple task even for the huge WWW since these iterations rapidly converge.[10] The PageRank of a page depends on the entire structure of the network, but, clearly, the PageRank depends strongly on the number of connections of a node. Typically, the higher the number of incoming hyperlinks of a page, the higher r_i. For uncorrelated networks, it is very easy to obtain from eqn (11.5) the average value $\overline{r}$ for a node with a given number of links [93]:

$$\overline{r}(q_{\text{in}}, q_{\text{out}}) = \frac{p}{N} + \frac{1-p}{N}\frac{q_{\text{in}}}{\langle q_{\text{in}} \rangle}. \tag{11.6}$$

Note that this average value is entirely determined by the in-degree of a node. This simple linear dependence describes the measured average PageRanks in the WWW surprisingly well, despite strong correlations in this network. In the range of high in-degrees, fluctuations of r_i from page to page with the same in-degree turn out to be relatively small. So in this range, the average value $\overline{r}(q_{\text{in}})$ approximates r_i extremely well. Equation (11.5) shows that to be highly ranked, a page must have incoming hyperlinks from pages with high PageRanks. Manipulating the connectivity of the nearest neighbours of a page, in principle, allows one to improve its PageRank. All the tricks, however, become virtually useless for pages with really high PageRanks.

[10] The solution of eqn (11.5) is actually the eigenvector of some matrix corresponding to its maximum eigenvalue. Finding these eigenvectors is among the simplest of numerical problems for matrices.

Traffic

12

The central function of a great number of networks is to provide convenient routes for the transportation of humans, objects, energy, information, and other items between their nodes. The goal is to obtain dense but free-flowing traffic. In this lecture we mainly discuss traffic and traffic congestion in complex networks.

12.1 Traffic in the Internet

The Internet is based on packet switching, which enables robust data transport.[1] In the process of transmission, (i) files are cut into packets; (ii) the packets are submitted (routed) 'independently' to their destinations through optimal routes; (iii) finally, at their destinations, the packets are reassembled into the original sequences. This technology allows networks to efficiently route information around failed nodes and links. In addition, packet switching distributes the flows of data over the network, and uses network resources fully. The header of each packet contains full addressing information, so that routers forward incoming packets by using this information and their routing tables. A routing table is the router's dynamically updated individual database of routes. Based on its routing table, a router forwards a packet to its neighbour which is expected to be on the optimal route to the destination.

[1] For a more detailed introduction to the basics of Internet traffic, see the paper of Willinger, Govindan, Jamin, Paxon, and Shenker [182] and a brief survey of concepts written by Smith [164].

The first theoretical studies of packet switching communications were made in early 1960s (Leonard Kleinrock, MIT), but the fundamental suite of communications protocols was written by V. G. Cerf and R. E. Kahn considerably later, between 1972–1974 [53]. These were the Transmission Control Protocol/Internet Protocol (TCP/IP). The basic protocol—IP—enables addressing and forwarding of packets, but IP is not sufficient for robust networking. TCP enables flow control and recovery from lost packets. The TCP/IP suite was adapted between 1980–1983.

Thus, traffic in the Internet is, in essence, a flow of packets from hosts–sources to hosts–destinations governed by Internet protocols. Of the rich set of features that this highly variable flow demonstrates, here we touch upon only two—self-similarity and fluctuations.

The self-similarity of Internet traffic is an incredibly widely discussed issue. It was believed until the first half of the 1990s that packet flow in the Internet is a *Poisson process*. This is a basic random process—a chain of uncorrelated events occurring at some average rate. For example, at each discrete moment, a random variable takes the value 0 with probability $1-p$ and the value 1 with probability p. Then the number of ones in some given time interval is distributed according to a Poisson distribution, and the number of zeros between two subsequent intervals is distributed according to an exponential distribution. If we coarsen this Poisson flow by binning—summing the ones within sufficiently long intervals of length T, then the resulting flow looks much smoother. The relative amplitude of fluctuations, $\sqrt{pT}/pT$, quickly decreases with T, and the flow is smoothed out. In 1993, Will Leland, Murad Taqqu, Walter Willinger, and Daniel Wilson thoroughly investigated the flow of packets collected from a network link and found that the real flow is in remarkably stark contrast to the Poisson process [118]. They discovered that Internet traffic viewed over different time scales of a few (in their case, four) orders of magnitude is highly bursty.[2] Over all these time scales, the flow $f(t)$ (the numbers of packets recorded during time intervals of a given length—'bins') looks 'self-similar' in the following sense. Scaling the time variable by a factor a rescales the flow:

[2] 'Bursty' means 'has a relatively large variance'.

$$f(t) = a^{-H} f(at)\,. \tag{12.1}$$

Here H is the so-called *Hurst exponent.* For Internet traffic, typically, $H \approx 0.8$. One can easily show that this self-similarity is equivalent to a slow power-law decay of the autocorrelation function:

$$\langle [f(t) - \langle f \rangle][f(t+\tau) - \langle f \rangle] \rangle \propto \tau^{-\tilde{\beta}}\,, \tag{12.2}$$

where exponent $\tilde{\beta}$ is related to the Hurst exponent: $\tilde{\beta} = 2(1-H)$, so $\tilde{\beta} \approx 0.4$.

What is the nature of this unexpectedly slow decay of correlations? The existing explanations reveal two contrasting views of the Internet—the network engineer's view and the physicist's one. The most widely accepted engineer's (computer scientist's) explanation is based on the broad distribution of users' sessions, for example, HTTP connections [182]. Empirical data demonstrate that the sizes of sessions (the numbers of packets in sessions or the lengths of transmitted files) are distributed according to a power law. This fact (supplied by an additional assumption that the sessions arrive uniformly at random) directly leads to the self-similar traffic of packets.[3]

[3] Exponent $\tilde{\beta}$ in eqn (12.2) is expressed in terms of the exponent $\tilde{\alpha}$ of the size distribution of sessions: $\tilde{\beta} = \tilde{\alpha} - 2$. Typically, $\tilde{\alpha} = 2.2$–2.4.

Physicists usually propose a quite different explanation. In their abstract models, they observe that on the verge of congestion, the flow of packets is power-law correlated, as is normal for a critical phenomenon. If a real network functions most efficiently, which corresponds to the highest possible traffic flow without congestion, then the traffic should be indeed critical and so, self-similar.

Computer scientists explain why the physicists' theories are not applicable to the Internet: 'self-similar scaling has been observed in networks with low, medium, or high loads' and not at some critical load [182].[4] The physicists' traffic models are too abstract and do not account for the real complexity of the Internet, ignoring, for example, even basic details of Internet protocols. This criticism pinpoints the common feature of

[4] We will return to this criticism in Lecture 14.

the physicists' theories of the Internet: physicists essentially ignore that the Internet is a 'highly engineered' system. They simply believe that this fact is not of primary importance. According to Reginald Smith—a network engineer, for physicists, 'the Internet is a complex dynamical system whose macroscopic properties are based on general mechanisms of statistical mechanics', while for network engineers, 'the Internet is a complex engineering system whose behaviour is determined by its protocols, users, and functions' [164]. We suggest, however, that in some form, the 'protocols, users, and functions' can be included into physicists' models. Why not?

Traffic variance (the amplitude of fluctuations) is an important characteristic of bursty traffic. Let the average flux measured at node i of a network be $\langle f_i \rangle$ and its standard deviation be $\sigma_i = \sqrt{\langle (f_i - \langle f_i \rangle)^2 \rangle}$. The question is: how are these quantities related? In 2004, Marcio de Menezes and László Barabási analysed data on traffic at a large number of nodes in the Internet and in a few other networks, which they collected over long time periods [68]. These empirical data were plotted as σ_i versus $\langle f_i \rangle$. It turned out that for the Internet, this set of points for different nodes could be fitted by a square root function: $\sigma \propto \langle f \rangle^{1/2}$. For traffic on the other investigated networks (for the number of visits of web sites, for the stream flow of rivers, for traffic on highways, and for flows within microprocessors), they found a proportional dependence $\sigma \propto \langle f \rangle$. The researchers suggested that the square root dependence is valid if flow fluctuations have an intrinsic nature, while proportional dependence is present if the variations of traffic are induced by some fluctuating external force. Later studies showed, however, that the empirical curves $\sigma_i(\langle f_i \rangle)$ are usually between these two limiting cases, the results depend on the time interval of the measurements, and so the situation is more complicated than was believed originally [124].

12.2 Congestion

The total traffic in the Internet grows even faster than the number of hosts, which in its turn grows exponentially with time. Traffic becomes more and more dense, which increases the risk of congestion, especially for highly connected nodes.[5] Let us discuss very basic features of this phenomenon. Suppose that a network has N nodes which all permanently send packets to each other, each node injects packets at a rate λ. If we assume that the routes are along the shortest paths between the nodes, then the flow through a node is proportional to the number of shortest paths which run through this node, or actually, to the betweenness centrality B_i of this node. The node with the highest betweenness centrality $B_{\max}$ will be the first congested. So it is easy to see that congestion emerges at the rate

$$\lambda_c = \frac{N-1}{B_{\max}}. \tag{12.3}$$

[5] On average, traffic increases by 50–60% every year [1].

This simple relation can be generalized to more general protocols which allow routes to depart from the shortest paths [168]. Importantly, the formula relates congestion and the architecture of a network. It is clear, in particular, that homogeneous networks have higher congestion thresholds than networks with hubs.

The natural method of delaying congestion, is to send packets around hubs. For example, we can use 'a random walk protocol', that is make all the packets random walkers. (Note, however, that this protocol is extremely inefficient because of long delivery times.[6]) Furthermore, we can increase the congestion threshold, by slightly modifying these random walks. Let a node forward a packet to its neighbour i chosen with probability proportional to q_i^μ [179,95]. Here q_i is the number of connections of this neighbour and μ is a given exponent for this random walk. Recall formula (12.1) for the stationary probability of finding a walker at a node of degree q for standard random walks: $p_{\text{fin}}(q) \propto q$. For the modified random walk, we similarly have $p_{\text{fin}}(q) \propto q^{1+\mu}$. The highest congestion threshold is apparently realized when p_{fin} is independent of q, which takes place at $\mu = -1$. Thorough calculations showed that this is indeed the case [179, 95].

[6] We do not discuss ways of making random walk protocols more efficient [175].

We can avoid congestion in another way. In *adaptive routing* protocols, instead of avoiding hubs, routed packets avoid nodes with long queues. In 2005, Pablo Echenique, Jesus Gómez-Gardeñes, and Yamir Moreno proposed a simple model of adaptive routing [84]. In this model, all nodes send packets to each other at a rate λ. Each node also functions as a router. Let n_i be the number of packets at node i at a given moment, which is a queue length. The buffers are assumed to be infinite. The nodes forward packets taking into account the queue lengths at their neighbours, but also trying to select the shortest routes to destinations. A packet is sent from a node to its neighbour j, which has the minimum value of the quantity:

$$\delta_j = h\ell_{jt} + (1-h)n_j \,. \tag{12.4}$$

Here n_j is the queue length at node j, ℓ_{jt} is the length of the shortest path from node j to the target t of the packet, and the parameter h of the model is a number between 0 and 1.

Quantitatively, congestion is characterized by the following order parameter [11]:

$$\rho = \frac{\text{\# of packets that failed to reach targets during the observation}}{\text{the total number of packets injected during this time period}} \tag{12.5}$$

or, equivalently, by

$$\rho = \frac{N_p(t+\Delta t) - N_p(t)}{N\lambda\Delta t} \,. \tag{12.6}$$

Here N is the network size, $N_p(t)$ is the number of packets in the network at time t, and Δt is the time of observation. In the jammed phase, the number of packets in a network grows linearly with time, which leads to time-independent ρ. Figure 12.1 shows the results of numerical simulations of this model, namely the dependencies of the order parameter on injection rate. The researchers used the map of a real Internet network. At $h=1$, the model provides shortest-path routing. In this limiting case, the transition to the congested phase was found to be continuous, without a jump. The curves become discontinuous at smaller values of h, and the congestion threshold significantly increases. Note that this transition resembles the birth of a k-core, compare Figs. 12.1 and 6.9 (c).

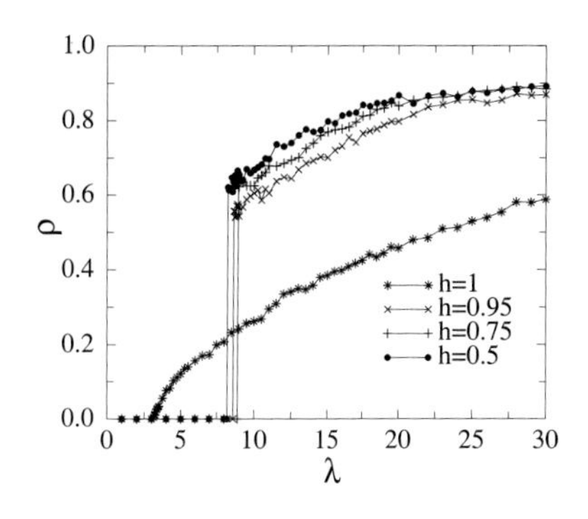

Fig. 12.1 Order parameter ρ, eqn (12.6), as a function of a packet injection rate λ in the Echenique–Gómez-Gardeñes–Moreno model of congestion. Adapted from [84].

12.3 Cascading failures

Wide-spread power outages in large electricity grids are among the most financially devastating events that occur in man-made systems. The latest large 2005 Java–Bali Blackout affected about 100 million people; the famous Northeast Blackout of 2003 affected 55 million people. The failure of a single element in a power grid triggers a cascade of failures across the entire grid or a large part of it. How can this happen? Each element in a general transportation system has a limiting load capacity. If a transportation network is fully or nearly fully loaded, then the failure of a single node or link overloads a few other elements, they fail, which leads to new overloads and failures, and so on.

In 2002, Adilson Motter and Ying-Cheng Lai proposed a simple model of these cascade processes [131]. In the Motter–Lai model, all nodes in a network permanently, at equal rate, send goods (or data packets, etc.) to each other along the shortest paths. Therefore the load of a node in this system is proportional to the number of shortest paths in the network that pass through this node. Recalling the definition of betweenness centrality, we can use this quantity as a load. Let $B_{0,i}$ be the betweenness centrality (load) of node i in an undamaged network. Motter and Lai assumed that each node has its limiting capacity, which is the maximum load the node will tolerate:

$$c_i = (1+\alpha)B_{0,i}\,. \tag{12.7}$$

Here $\alpha \geq 0$ is a 'tolerance parameter', which shows how much an initial load can be exceeded without failure. If the betweenness centrality b_i of a node exceeds its capacity c_i, this node is immediately removed. A cascading failure in this model occurs in the following way.

(i) Compute the betweenness centralities $B_{0,i}$ and so capacities c_i of all nodes in a network.

(ii) Remove a chosen node from the network.

(iii) Compute all the betweenness centralities $B_{1,i}$ in the resulting network.

(iv) Delete all the overloaded nodes where $B_{1,i} > c_i$.

(v) Repeat (iii)–(iv) until no overloaded nodes remain.

[7] Note that there is no self-averaging here. For each of the three types of initial failures, we have to average over all starters. In the simulations, it was sufficient to average over a few triggers.

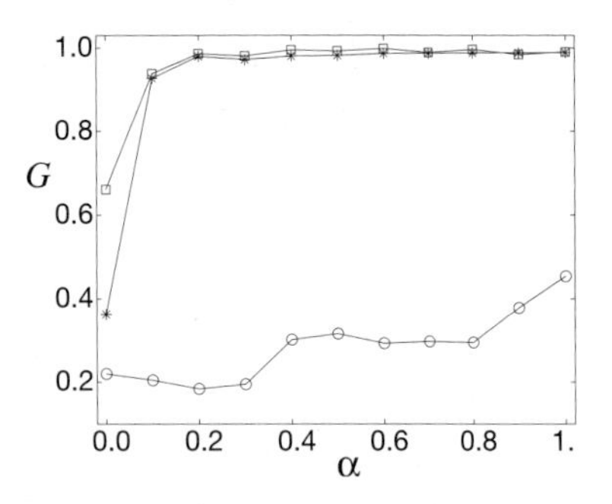

Fig. 12.2 Modelled cascading failure in the Western US power transmission grid (4941 nodes with, on average, 2.67 connections). The curves show the relative size of a giant connected component remaining after the cascading failure versus the tolerance parameter α. The cascades are triggered by the removal of a single node selected: (i) uniformly at random ($\square$), (ii) among nodes of highest degrees ($*$), and (iii) among nodes with highest loads ($\circ$). Adapted from Motter and Lai [131].

[8] These were recursive scale-free networks with degree-distribution exponent $2 < \gamma \leq 3$. The triggers were chosen uniformly at random.

In numerical simulations, Motter and Lai studied a number of loopy undirected networks. The main quantity of interest was the relative size of a giant connected component that remains in a network after a cascading failure: $G = N'/N$, where N' is the size of the resulting giant component, and N is the total number of nodes in the original network before the failure. The researchers, in particular, investigated the dependence of this quantity on the tolerance parameter α. It is clear that zero α guarantees the complete collapse of a network, so the question is: how rapidly does G increase with α? Note that the size of a cascade failure depends significantly on a starting node. Figure 12.2 shows three distinct curves $G(\alpha)$ for three kinds of triggers—first failed nodes: for a node selected uniformly at random, for a node chosen from among the most connected nodes, and, finally, for a node chosen from among the most loaded nodes in an original network.[7] The figure shows that, as is natural, the largest cascades are triggered by failed nodes with the highest loads. Later, comprehensive numerical simulations of large networks uncovered more detailed features of cascading failures [117].[8] It has been found that there is a threshold tolerance, $\alpha_c > 0$, above which all cascading failures are finite, $G = 1$ (compare with the curve for uniformly random triggers in Fig. 12.2 where $G \approx 1$ for $\alpha > 0.2$). Importantly, at this threshold, the size distribution of cascades is power-law, which shows that this point is critical. That is, the Motter–Lai model has a phase transition to the state with giant avalanches.

Interacting systems on networks

13

In this lecture we consider systems of interacting agents placed on networks. The agents—spins, oscillators, interacting individuals, etc.—occupy the nodes of networks and interact with each other through network links. In more complicated situations, these agents, in turn, influence their network substrates, and so the pair—a network and the system of agents—co-evolve. We discuss unusual phenomena in these cooperative systems on complex networks.

13.1 The Ising model on networks

In 1924 the young scientist Ernest Ising (1900–1988) received a Ph.D. in physics from the University of Hamburg. Ising's Ph.D. thesis contained a study of the basic model of a ferromagnet, proposed to Ising by his renowned supervisor, Wilhelm Lenz (1888–1957). This famous reference model—now known as the Ising model—will help us to understand the main features of interacting systems placed on various substrates. The Ising model describes interactions between classical spins ($s_i = \pm 1$) in an idealized ferromagnet. The energy of a given configuration of spins in this model (physicists call it Hamiltonian) is

$$\mathcal{H} = -\sum_{i,j} J_{ij} s_i s_j - H \sum_i s_i \,, \tag{13.1}$$

where the sum is over all pairs of nearest-neighbouring spins, and H is an applied magnetic field. The couplings $J_{i,j}$ are positive, so the 2-degenerate ground state of the model is ferromagnetic: all spins in the system are of the same sign—either plus or minus. When the thermal energy $k_B T$ exceeds the energy of coupling of a spin to its z nearest neighbours, $z\overline{J}$, this ferromagnetic order is destroyed. Here, $\overline{J}$ is the mean coupling. The system enters the paramagnetic phase, where each spin frequently flips, so that the average value of each spin $\langle S_i \rangle = M_i$ and the macroscopic magnetic moment are zero. The critical temperature, separating ferromagnetic and paramagnetic phases is estimated as $T_c \sim z\overline{J}/k_B$, where k_B is the Boltzmann constant.

Let us start from the Ising model on the simplest infinite-dimensional system—on the fully connected graph of N nodes, where N approaches infinity. For simplicity, assume that no external field is present. Thermal fluctuations force each spin to flip from time to time and result in

$|M_i(T)| < 1$. Variations in the average values of spins $\langle S_i \rangle = M_i(T)$ with temperature essentially characterize the cooperative behaviour of the system. We can easily obtain equations for $M_i(T)$ using a mean-field approach. Notice that each spin i is in effect in the field $H_i = \sum_{j \neq i} J_{ij} s_j \approx \sum_{j \neq i} J_{ij} M_j$. Here the second equality is what is called the mean-field approximation. In our system, thanks to the infinite number of nearest neighbours of a spin, this approximation is apparently exact. As a result, we obtain the equations for $M_i(T)$:

$$M_i = \frac{e^{\beta H_i} - e^{-\beta H_i}}{e^{\beta H_i} + e^{-\beta H_i}} = \tanh\left[\beta \sum_{j \neq i} J_{ij} M_j\right], \tag{13.2}$$

where $\beta = 1/k_B T$. These equations can easily be solved. For a homogeneous ferromagnetic system, we set $J_{ij} = J/N > 0$ to get a finite critical temperature in the limit $N \to \infty$. In this case, we easily find the critical temperature $T_c = J/k_B$ and the relative macroscopic magnetic moment $M(T) = \sum_i M_i(T)/N$ near the critical point:

$$M(T) \propto \sqrt{T_c - T}\,. \tag{13.3}$$

Singularities of this kind correspond to continuous phase transitions. This square root is one of two possible critical peculiarities for the order parameter in traditional mean-field theories of second-order phase transitions.[1]

[1] The second one was discussed in Lecture 6: for percolation problems, the traditional mean field theory gives the size of a percolation cluster $S \propto p - p_c$.

The same critical singularity and a similar critical temperature were found in the ferromagnetic Ising model on sparse classical random graphs. What about complex networks? We can apply eqn (13.3) even to these complicated situations. Our hope is that mean-field theories should work well for infinite-dimensional systems. Consider the ferromagnetic Ising model on top of a sparse uncorrelated graph with a given degree sequence $q_1, q_2, \ldots, q_N$. Make the following *annealed network approximation* [26]. Substitute the original random network with a fully weighted connected graph with link weights $q_i q_j / \langle q \rangle N$. In other words, substitute the original network with an effective medium, where the sum of couplings for each node i has the same value Jq_i as in the original graph. Equivalently, force the links of the network to jump frequently between different pairs of nodes while keeping a given set of values Jq_i. 'Frequently' here means that links jump more frequently than spins flip. In that sense, the set of links is 'annealed'. This approximation provides a surprisingly good description of cooperative models on sparse uncorrelated networks. In this lecture we focus on these complex networks. For the resulting fully connected non-random graph, eqn (13.2) gives

$$M_i = \tanh\left[\frac{\beta J q_i}{\langle q \rangle N} \sum_j q_j M_j\right] = \tanh[\beta J q_i \mathcal{M}]\,. \tag{13.4}$$

A weighted magnetic moment $\mathcal{M} = \sum_j q_j M_j / \langle q \rangle N$ is a solution of the following equation:

$$\mathcal{M} = \sum_q \frac{q P(q)}{\langle q \rangle} \tanh[\beta J q \mathcal{M}]\,. \tag{13.5}$$

Here $P(q)$ is the degree distribution of the original random network. Interestingly, it is not M but rather the weighted moment $\mathcal{M}$ that plays the role of the order parameter in this problem. In simple terms, $J\mathcal{M}$ is an effective field acting on a spin from its nearest neighbour in this network. To obtain eqn (13.5), substitute the expression for M_i from eqn (13.4) into the definition of $\mathcal{M}$ and use the equality $\sum_i f(q_i) = \sum_q P(q) f(q)$. To find how average spins and the macroscopic magnetic moment depend on temperature, we should first solve eqn (13.5) and then substitute the solution $\mathcal{M}$ into eqn (13.4).

Equation (13.5) is a key point in our problem. Recall a similar eqn (6.2) for the order parameter x for the emergence of a giant connected component in uncorrelated networks. Importantly, in both of these equations, the nearest-neighbour degree distribution $qP(q)/\langle q \rangle$ appears in the sum on the right-hand side. We explained this factor for the giant connected component problem. Moreover, for arbitrary pairwise interactions, the description should apparently involve the degree distributions for the ends of a link, the same $qP(q)/\langle q \rangle$, as in eqn (13.5). The presence of this factor crucially increases the role of highly connected nodes in these systems, especially for various critical phenomena.

13.2 Critical phenomena

From eqn (13.5), we readily find the critical temperature,

$$k_B T_c = J \frac{\langle q^2 \rangle}{\langle q \rangle} \,. \tag{13.6}$$

This formula is approximate. However, agreement with a rigorously obtained expression is more than satisfactory.[2] Consequently, T_c approaches infinity, if the second moment of the degree distribution diverges. In particular, for scale-free networks, thermal fluctuations cannot destroy the ferromagnetic order at any finite temperature if the degree distribution exponent γ is equal to or less than 3 [7]. In other words, hubs support cooperative ordering in these systems. This infinite critical temperature in a cooperative model on a sparse network is in a sharp contrast to what we know for lattices. This is similar to percolation—hubs prevent the destruction of a giant connected component by randomly damaging a network.

[2] This approximate result is close to the exact value

$$k_B T_c = \frac{2J}{\ln[\langle q^2 \rangle / (\langle q^2 \rangle - 2\langle q \rangle)]}$$

obtained for uncorrelated networks (see, Refs. [75, 77]).

When the fourth moment $\langle q^4 \rangle$ is finite (that is $\gamma > 5$), expand the hyperbolic tangent $\tanh[\beta J q \mathcal{M}] \approx \beta J q \mathcal{M} - (\beta J q \mathcal{M})^3/3$ in the sum on the right-hand side of eqn (13.5). The resulting equation readily gives the same square-root critical singularity (13.3) as the classical random graphs and lattices. A new, non-traditional kind of critical singularity emerges if exponent $3 < \gamma \leq 5$. For these heavy-tailed degree distributions, we cannot expand $\tanh[\beta J q \mathcal{M}]$ in eqn (13.5). Instead, we do the following. We add and subtract $\mathcal{M} T_c / T = \mathcal{M} \beta J \langle q^2 \rangle / \langle q \rangle$ from the right-hand side of this equation:

$$\mathcal{M} = \mathcal{M} \sum_q \frac{qP(q)}{\langle q \rangle} \beta J q + \mathcal{M} \sum_q \frac{qP(q)}{\langle q \rangle} \frac{\tanh[\beta J q \mathcal{M}] - \beta J q \mathcal{M}}{\mathcal{M}} . \quad (13.7)$$

The second sum on the right-hand side in this equation is a singular function of $\mathcal{M}$. Indeed, we can estimate the sum as follows:

$$\beta J \frac{\langle q^2 \rangle}{\langle q \rangle} - 1 = \sum_{q \sim 1/\mathcal{M}}^{\infty} \frac{qP(q)}{\langle q \rangle} \beta J q \propto \int_{1/\mathcal{M}}^{\infty} dq \, q^{2-\gamma} \propto \mathcal{M}^{\gamma-3} , \quad (13.8)$$

and so the critical singularity is

$$M \propto (T_c - T)^{1/(\gamma-3)} \quad (13.9)$$

when $3 < \gamma < 5$.[3] Recall a similar critical behaviour for percolation on complex networks. The singular dependence (13.9) is actually exact [75,119]. Remarkably, we obtained it using simple mean-field equations. In principle, the exactness of the mean-field approach is not a surprise since these networks are infinite-dimensional. The real surprise is a dramatic change of the critical behaviour in networks with heavy-tailed degree distributions. For sufficiently homogeneous substrates (exponent $\gamma > 5$), we observe a standard square root critical singularity. If a network substrate for the cooperative system is strongly heterogeneous ($\gamma < 5$), we find non-traditional singularity.

[3] In the critical region, $M \propto \mathcal{M}$. Note that the average magnetic moment of a node increases with the number of its connections, see eqn (13.4). For hubs, this moment approaches 1 even near the phase transition.

Physicists know that in the Ising model on two- and three-dimensional lattices, the critical exponent of the order parameter also differs from the mean-field value 1/2, but is smaller than this number. According to the standard classification, this range of exponent values corresponds to second-order phase transitions. In contrast, for the Ising model on scale-free networks, the exponent increases from 1/2 to ∞ as the degree distribution exponent decreases from 5 to 3. This indicates a progressive increase of the phase transition order from second to infinite.

Strong structural correlations in a network may change this picture. Moreover, on many networks, the Ising model has no ferromagnetic ordering at all. For example, the Ising model on any tree—a network without loops—is ordered only at zero temperature. Even weak thermal fluctuations break this ordering. Similarly, eliminating a tiny fraction of nodes in a tree, we split it into a set of finite components. On the other hand, we can easily restore macroscopic ordering in these systems by adding even a relatively small number of shortcuts between randomly chosen nodes.

13.3 Synchronization

What is synchronization?

The phase transitions, which we considered in preceding sections occur only in infinite (or, in reality, very large) interacting systems. At any temperature, in a finite spin system, thermal fluctuations flip the magnetization from time to time, and on average, the magnetic moment is zero. There is, however, a wide circle of phenomena, where even a small set of interacting units, e.g. oscillators, can enter a coherent (synchronous) state. This synchronization is possible even for a pair of coupled oscillators. A traditional example is two pendulum clocks mounted on the same wall, which swing in unison. Synchronization occurs in a wide range of dynamical systems of very different sizes. Recall swarms of synchronously flashing fireflies, cicadas, and crickets chirping in unison and so on. Moreover, it is even possible that a finite dynamical system demonstrates synchronization, but its infinite counterpart does not.

The reader will find a detailed discussion of diverse synchronization phenomena in various networks in review [12]. Many real-life systems exhibiting synchronization have really complex architectures. A wide spectrum of synchronization phenomena, for example, was observed in complex networks in the brain. The brain contains a hierarchy of networks. At the cellular level, these are neuronal networks, where waves of activation spread over neurons (activators and inhibitors) interconnected by directed links—synaptic connections.[4] At a larger scale, researchers consider weighted cortical networks, in which nodes are interconnected cortical areas. Temporal patterns of brain electrical activity show a complicated set of oscillations at various frequencies, mostly in the range 1–100 Hz. Each pattern of rhythmic activity is closely related to a specific regime of brain functioning. Scientists are still far from understanding these brain rhythms and similar oscillatory behaviours of traffic and network flows in the Internet, cellular and ecological networks. Only for very simple dynamic models could researchers find a relation between the quality of synchronization and network architecture. We will discuss two of the basic systems exhibiting synchronization.

[4] Each directed weighted link in this network is a chain: axon–synapse(s)–dendrites, where synapses can change their states recording information. The human brain contains about 10^{11} neurons and 10^{14} synapses.

What is phase synchronization in essence?[5] The phase $\theta(t)$ of a single phase oscillator with an angular frequency ω evolves linearly, $\theta = \omega t + \text{const}$. This evolution is described by a trivial differential equation, $\dot{\theta}(t) = \omega$. For a pair of independent oscillators with two different frequencies ω_1 and ω_2, we have two independent equations: $\dot{\theta}_1 = \omega_1$ and $\dot{\theta}_2 = \omega_2$. Suppose that these oscillators are coupled in some way. Then, when is it possible that θ_1 and θ_2 evolve in phase, despite the difference between ω_1 and ω_2? Here 'in phase' means that the difference between the phases of the oscillators approaches a constant value. To couple oscillators, let us add new terms to the right-hand sides of the motion equations for the phases. These interaction terms must be zero when the phases coincide (mod2π, of course) and they must increase as the phases deviate (mod2π). The simplest relevant periodic function is a sinusoid. So we arrive at a pair of coupled equations,

[5] See the popular science book of Steven Strogatz [171] for a wonderful introduction to the topic.

$$\dot{\theta}_1 = \omega_1 + J\sin(\theta_2 - \theta_1)\,,$$
$$\dot{\theta}_2 = \omega_2 + J\sin(\theta_1 - \theta_2)\,, \qquad (13.10)$$

where J is a coupling constant. For the difference $\varphi = \theta_1 - \theta_2$, these equations give

$$\dot{\varphi} = \omega_1 - \omega_2 - 2J \sin\varphi\,. \tag{13.11}$$

Synchronization means $\varphi(t \to \infty) \to$ const. For this, the equation $\omega_1 - \omega_2 = 2J \sin\varphi$ must have a solution. Consequently, the oscillators synchronize if the coupling is sufficiently strong, namely, $J > |\omega_1 - \omega_2|/2$. For smaller J, oscillators run incoherently.

Oscillators with random frequencies

Yoshiki Kuramoto at Kyoto University made a major breakthrough in the 1970s. He found an exact solution for the direct generalization of this model, eqn (13.10), to the case of N coupled oscillators with natural frequencies ω_i, $i = 1, 2, \ldots, N$ [116]. Kuramoto solved the model on the infinite fully connected graph, but, generally, the *Kuramoto model* is formulated on an arbitrary graph:

$$\dot{\theta}_i = \omega_i + J \sum_{j=1}^{N} a_{ij} \sin(\theta_j - \theta_i)\,. \tag{13.12}$$

Here we assume that interactions between oscillators are through the links of a network with an adjacency matrix a_{ij}.[6] The frequencies ω_i in the Kuramoto model are random numbers from some distribution function, say, of a bell-shaped form, $\overline{\omega} = \sum_i \omega_i / N$. When $N > 2$, it is hard to follow the evolution of many phases $\theta_i(t)$, and synchronization becomes essentially more complicated and interesting than for two oscillators. The reader can suggest that when $N > 2$, only some fraction of the oscillators may be in phase, that is, the differences between their phases are constant, while the rest run incoherently. It turns out that this is indeed the case. To describe the collective dynamics of many individual phases and their mutual entrainment, Kuramoto used *the complex order parameter*:

$$r(t)e^{i\psi(t)} = \frac{1}{N} \sum_{k=1}^{N} e^{i\theta_k(t)}\,. \tag{13.13}$$

Here $\psi(t)$ plays the role of the average phase, and the module $r(t)$ characterizes the extent of coherence.

According to Kuramoto, the picture looks as follows, see Fig. 13.1. If interactions are sufficiently weak, the system stays in an incoherent state. In this state, the phases $\theta_i(t)$ drift more or less at random on the unit circle in the complex plane. They are homogeneously scattered over the circle, and so $r = 0$, see Fig. 13.1(a). There is a critical value of coupling constant above which this picture changes sharply. This critical coupling constant is of the order of the width of the distribution of oscillator frequencies, $\delta\omega$. Above the critical point, the phases are partially synchronized. Namely, a fraction of phases are coherent, which are locked and scattered around the average phase $\psi(t)$. These locked oscillators oscillate with the same frequency, see Fig. 13.1(a). (Their natural frequencies are in the region around $\overline{\omega}$.) The phases of the remaining oscillators drift incoherently. As interactions become stronger,

[6] For an infinite fully connected graph, $a_{ij} = 1$ and J is substituted for J/N.

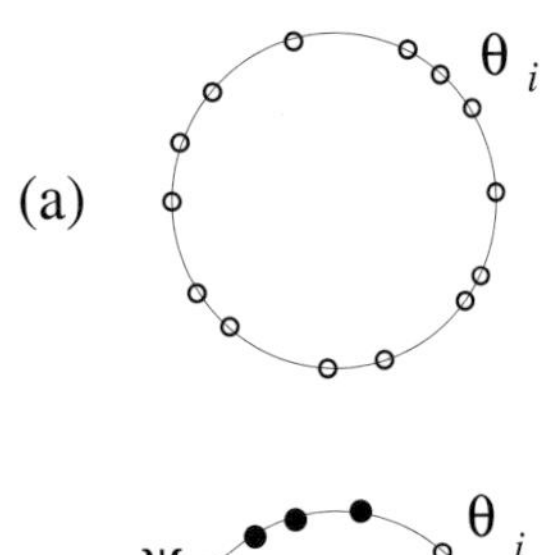

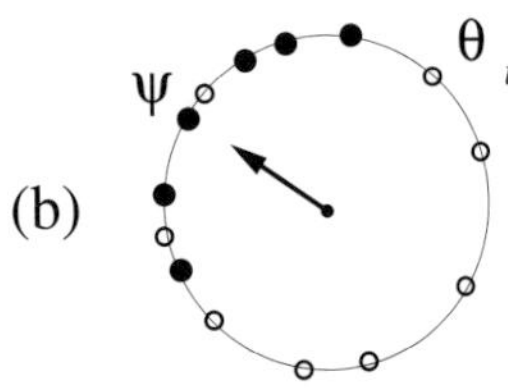

Fig. 13.1 Snapshots of the phases $\theta_i(t)$ of oscillators. (a) The incoherent state. The open dots show the drifting phases of oscillators on the unit circle in the complex plane. (b) The synchronized state. The filled dots show coherent (or locked) phases. The remaining oscillators drift incoherently. The direction of the vector and its length show the average phase $\psi(t)$ of the oscillators and the module of the complex order parameter, respectively. For more details, see the introductory review of Strogatz [172].

the number of locked oscillators grows, and, in addition, their phases draw nearer to the average phase. Consequently, the module of order parameter, r, increases.

In the Kuramoto model on the infinite fully connected graph, r is proportional to the square root of the deviation of the coupling constant from its critical value, $\sqrt{J - J_c}$. The singularity is the same as for a ferromagnetic second-order phase transition. On the other hand, in sharp contrast to ferromagnets, the Kuramoto model on infinite finite-dimensional lattices does not show synchronization at any coupling strength.[7]

[7] Here the distribution of natural frequencies is supposed to rapidly decrease. Long-range shortcuts restore synchronization in these infinite systems.

Instead of the fully connected graph, as a substrate for the Kuramoto model, we could use a classical graph. The results are similar. What about more complex and realistic networks? Suppose, a network is sparse and uncorrelated. For this network, one can use an effective medium approach similar to the one that we described for the Ising model [104]. This theory shows that for synchronization, the coupling constant must exceed the critical value

$$J_c \sim \delta\omega \langle q \rangle / \langle q^2 \rangle \,. \tag{13.14}$$

Here $\delta\omega$ is the width of the distribution of oscillator frequencies. Therefore, if the degree distribution of a network is heavy-tailed then there is synchronization even at a very small interaction strength. In that sense, hubs in a network significantly improve the synchronization of Kuramoto oscillators. On the other hand, when interaction J is weak, only a small fraction of the oscillators turn out to be synchronized. In this case, the cluster of coherent oscillators includes the hubs and, mostly, their neighbours. The cluster gradually increases with J.

This scenario is similar to what takes place in percolation problems on this network and what we observed for the Ising model, and so these theoretical predictions look natural. Unfortunately, it was hard to verify these predictions in time-consuming numerical simulations. The maximal sizes of studied networks did not exceed 10^5 nodes. These sizes, however, were sufficient to observe a surprisingly strong deviation from the results of the effective medium approach. Yamir Moreno and Amalio Pacheco simulated the Kuramoto model on the Barabási–Albert network [130] and found an unexpected synchronization transition. In the version of the Barabási–Albert model, that they used, each new node had several connections, so that the network was loopy. In this network, the second moment $\langle q^2 \rangle$ diverges (the degree distribution decays as q^{-3}). So, in accordance with eqn (13.14), one could expect partial synchronization at any coupling constant. Nonetheless, Moreno and Pacheco observed that the critical coupling constant is not zero, $J_c > 0$, and below this value any coherence is absent.

The first idea was that the observed transition could be a finite size effect. Indeed, the second moment of the degree distribution slowly varies with network size N, $\langle q^2 \rangle \sim \ln N$.[8] Consequently, if eqn (13.14) works for this network, then $J_c \propto 1/\ln N$. The logarithmic function is so slow that this estimate gives a noticeable J_c even if the network is very

[8] In this network, the cut-off of the degree distribution is $q_{\text{cut}} \sim \sqrt{N}$, so

$$\langle q^2 \rangle \sim \int^{q_{\text{cut}}} dq\, q^2 q^{-3} \sim \ln q_{\text{cut}} \sim \ln N.$$

large. For example, $1/\ln 10^4 \approx 0.11$, $1/\ln 10^6 \approx 0.07$, $1/\ln 10^8 \approx 0.05$. In situations of this kind, researchers study a set of networks of various sizes. Moreno and Pacheco investigated the networks of 10^3, 10^4, and 5×10^4 nodes. For each of these networks they measured the variation of the order parameter r with interaction strength J. All of these curves were rather smooth, but it was possible to evaluate the dependence $J_c(N)$. It turned out that J_c does not decrease with N in contrast to our estimate. On the contrary, $J_c(N)$ increases, apparently approaching some constant value. Then, why does formula (13.14) fail in this case? At present nobody can give a definitive answer. The failure is probably because the growing Barabási–Albert network significantly differs from the uncorrelated networks.

Identical oscillators

There is a wide class of synchronization systems which are fundamentally different from the Kuramoto model. In these systems, interacting oscillators are identical. This is why the synchronization is not partial, as in the Kuramoto model, but full. So the structure of the coherent state is quite definite and clear. The natural question then is: what are the conditions of this synchronization? Namely, what is the range of coupling constant values and network architectures that provide this synchronization?

Physicists formulate models for this kind of synchronization in a very general form. The state of an individual oscillator is described by some vector $\mathbf{x}$. The equation of motion is the same for each oscillator, $\dot{\mathbf{x}} = \mathbf{F}(\mathbf{x})$, when the oscillators are uncoupled. Here $\mathbf{F}(\mathbf{x})$ is a given vector function. It is easy to make these oscillators interacting. For this, we couple the equations of motion for the oscillators, $i = 1, 2, \ldots, N$, together in the following way:

$$\dot{\mathbf{x}}_i = \mathbf{F}(\mathbf{x}_i) + J \sum_j a_{ij}[\mathbf{H}(\mathbf{x}_j) - \mathbf{H}(\mathbf{x}_i)] \,. \tag{13.15}$$

Here $\mathbf{H}(\mathbf{x})$ is a so called 'output function' and a_{ij} are the elements of the adjacency matrix of the underlining network. The functions $\mathbf{F}(\mathbf{x})$ and $\mathbf{H}(\mathbf{x})$ are given characteristics of the model.[9] The sum on the right-hand side of eqn (13.15) is usually represented in another form, namely

$$\sum_j (q_i \delta_{ij} - a_{ij}) \mathbf{H}(\mathbf{x}_j) \,,$$

where q_i is the degree of node i. In mathematics, the matrix $q_i\delta_{ij} - a_{ij}$ is called Laplacian. Therefore, to evaluate the influence of the network architecture on synchronization, we should study the Laplacian matrix of a given network. These matrices play a paramount role in graph theory and have numerous applications. The spectra of Laplacian matrices (Laplacian spectra) determine the dynamics of various processes on networks, for example, a random walk of a particle and, more generally, a wide range of diffusion processes. Furthermore, we do not need to

[9] Note that eqn (13.15) and corresponding eqn (13.10) in the Kuramoto model have similar structures, although the methods of coupling the oscillators are different. Compare the terms: $\sin(\theta_j - \theta_i)$ and $\mathbf{H}(\mathbf{x}_j) - \mathbf{H}(\mathbf{x}_i)$.

know a full Laplacian spectrum, see Fig. 13.2. Two physicists, Mauricio Barahona and Louis Pecora showed that the presence of this specific synchronization is determined by only two numbers in the Laplacian spectrum of a network—the minimum non-zero eigenvalue and the maximum one [17].

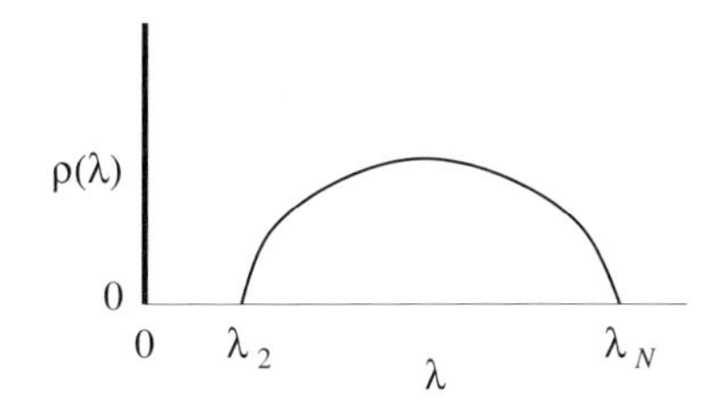

Fig. 13.2 Typical Laplacian spectrum of a large network. Let a graph of N nodes consist of a single connected component. Its Laplacian spectrum has N eigenvalues: λ_1=0 (see the peak at λ=0 in the plot) and $0<\lambda_2<\ldots<\lambda_N$. In the fully connected graph, $\lambda_2 = \ldots = \lambda_N$. In an infinite lattice, $\lambda_2 \to 0$.

According to Barahona and Pecora, a network is synchronizable (in other words, its synchronization state is stable) if the coupling strength J is in the interval $(\alpha_1/\lambda_2, \alpha_2/\lambda_N)$. Here the numbers α_1 and α_2 depend only on the specific functions $\mathbf{F}(\mathbf{x})$ and $\mathbf{H}(\mathbf{x})$. These numbers are positive, $0 < \alpha_1 < \alpha_2$. Typically, α_2/α_1 is in the range from 5 to 100. Consequently a network is synchronizable if

$$\frac{\lambda_N}{\lambda_2} < \frac{\alpha_2}{\alpha_1}. \tag{13.16}$$

The left-hand side is determined by the architecture of the network, while the right-hand side is determined by the nature of the oscillators and their interactions. When we consider the role of the network structure, we treat the ratio α_2/α_1 as a given parameter in the problem. Thus the smaller the ratio λ_N/λ_2, the better is synchronizability.

Criterion (13.16) is applicable to any network and lattice. In the infinite lattices, there is no gap between the first eigenvalue and second, that is $\lambda_2 = 0$. So we readily see that synchronization is absent at any coupling strength if a lattice is infinite. In the finite lattices, $\lambda_2 > 0$, and synchronization is possible. It was found that in the small-world networks, shortcuts diminish the ratio λ_N/λ_2, see Fig. 13.3, and so with shortcuts, infinite lattices become synchronizable. The same is true for phase synchronization in the Kuramoto model. Thus in both classes of interacting systems which we have discussed, the transition from a lattice to a small world results in synchronization. In the beginning of this section we listed a number of real-world systems exhibiting synchronization. Some of them are large. It is the small-world topology of these systems that makes synchronization possible.

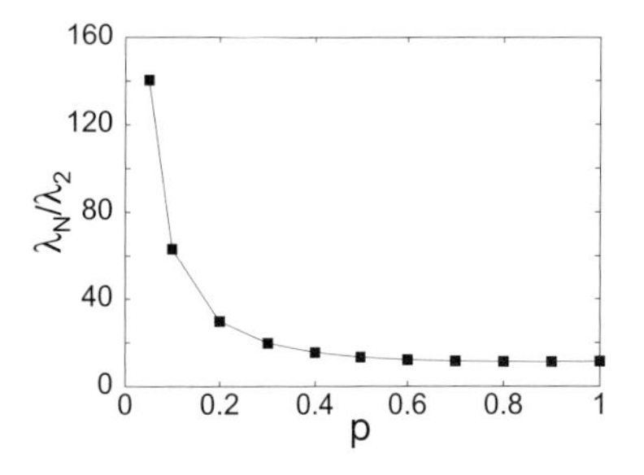

Fig. 13.3 Dependence of the ratio λ_N/λ_2 on the fraction p of shortcuts for the infinite Watts-Strogatz network. Adapted from Hong, Kim, Choi, and Park [103].

Criterion (13.16) shows that the fully connected graph ($\lambda_2 = \lambda_N$) provides the best synchronizability. For a sparse network with a given mean degree of a node, the narrowest Laplacian spectrum is for a random regular graph and then for a classical random graph. The minimum ratio λ_N/λ_2 provides the best synchronizability. For networks with heavy-tailed degree distributions, we have wide Laplacian spectra. It turns out that λ_N is determined by the cut-off of a degree distribution, $q_{\text{cut}}(N)$, while λ_2 is independent of the cut-off. In particular, for uncorrelated networks, $\lambda_N = q_{\text{cut}} + 1$ [108]. This crucially impairs the synchronizability of scale-free networks.

It is possible to arrive at the same conclusions in another way. Donetti, Hurtado, and Muñoz addressed the problem: what is the best synchronizable network for given numbers of nodes and links? To find the answer, they used a numerical optimization procedure [72]. They started from an arbitrary connected graph, having given numbers of nodes and links, and successively rewired uniformly randomly chosen links to randomly chosen nodes. For each rewiring, they checked whether it improved synchronizability. If it improves, then rewire; if it impairs, then abort this rewiring and try the next one. This time-consuming optimization allowed them to study only very small networks of a few hundred nodes. All the resulting optimal networks turned out to be cage graphs, see Fig. 1.2, that is networks with all nodes of equal degree, having loops (cycles) of the maximal possible length. Among non-random graphs of a given size, cage graphs are the closest to random regular graphs, so, indeed, homogeneous architectures provide the best synchronizability.

13.4 Games on networks

Let us pass now from interacting spins and oscillators on networks to specifically interacting individuals, namely, to players. The game which we choose for our players is the famous ***prisoner's dilemma***, which is one of the standard problems in game theory. Each of two players ('prisoners') independently decides for himself which is better: to cooperate (remain silent), C, or to defect (betray), D? This decision should be based on the following set of payoffs for the players:

Players:	Their payoffs:
(C, C)	$(1, 1)$
(C, D)	$(0, b)$
(D, C)	$(b, 0)$
(D, D)	$(0, 0)$

Here we present the simplest non-trivial set. Importantly, b is assumed to be greater than 1. If both players decide to cooperate, then they will receive equal payoffs, 1. If the first player cooperates, and the second defects, then the first will get nothing, 0, while the second will receive b, and so on. Then, what is better: to cooperate or to defect? What is more profitable in terms of payoff? In this simple game, a strategy providing the highest payoff is obvious. Since $b > 1$, the best action for a player is to defect. Indeed, suppose first that his opponent defects. Then, irrespective of the strategy of the first player, he will receive zero payoff. On the other hand, if the second player cooperates, then the first will get a bigger payoff by defecting, b against 1. Then it is better to always defect. The paradox is that if both the players use the defector strategy, then, together, they score less than cooperating: $0+0 < 1+1$.

The game becomes much more interesting when players can adopt the strategies of their opponents, and when there are many players [173]. For a network of players, at a given moment, each node (player) i can be in one of two states, $\sigma_i = C$ or $\sigma_i = D$, depending on the strategy that the player uses at this moment. Each pair of nearest neighbours independently play the game and receive their payoffs determined by the states of these pairs, see the list above. To make this system evolve we must allow individual players to change their strategies from time to time, $\sigma_i{=}C \longleftrightarrow \sigma_i{=}D$. Suppose that the players have information only about the results of their opponents—neighbours. Then a natural idea for an adaptive player is to adopt the most successful strategy in his close environment (the player himself and his nearest neighbours).

This idea was realized in the so-called *spatial prisoner's dilemma* of Martin Nowak and Robert May [142]. Without going into detail, in this mathematical model, originally formulated for a regular graph, the evolution of the players' strategies appears as follows. Initially, there is a random configuration of defectors and cooperators. For example, the initial concentration of cooperators c may be set to $1/2$. Then all of the pairs of nearest neighbours play the game independently, and after this round, each player i accumulates his payoffs as P_i. After that, for each player i, choose at random one of its immediate neighbours, j, and compare the scores P_i and P_j.[10] If $P_i > P_j$, then leave the state (strategy) of i unchanged. Otherwise, let player i accept the strategy of j with some probability. After this update, pass to the next round, recalculate all the scores and so on. As a result of this evolution, the concentration c of cooperators approaches a stationary value. This value depends on the parameter b of the prisoner's dilemma.

[10] The original version of the model is deterministic: each player adopts the best strategy in his close environment [142]. Here we describe a stochastic version.

This evolutionary model was generalized to other, non-regular networks and simulated [158]. Figure 13.4 shows the resulting final concentrations of cooperators in contrasting networks (regular graphs and Barabási–Albert networks). The reader can see that scale-free architectures stimulate and support cooperation. It turns out that players occupying the hubs tend to be cooperators, which makes cooperation dominant in these networks. These observations may, at least partially, explain why cooperation is widespread in heterogeneous populations. One can go even further and couple two evolutionary processes in these systems, namely, the evolutions of players and of a network [146]. In this generalization, not only does the network influence the course of the game, but also the dynamics of the game changes the structure of the network.

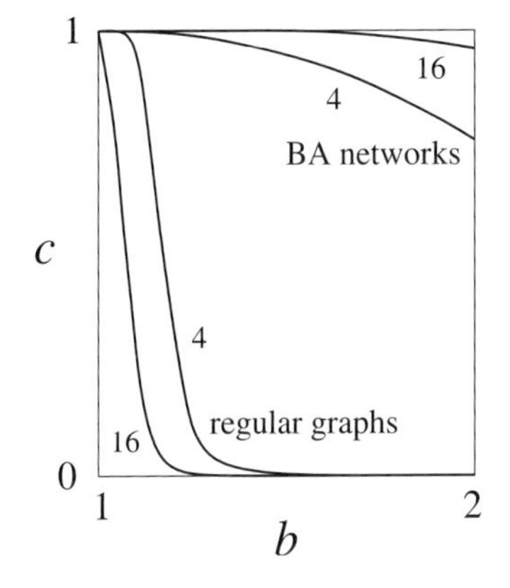

Fig. 13.4 Schematic plot of a final concentration c of cooperators versus parameter b for two contrasting networks according to Santos and Pacheco [158]. The numbers labelling the curves indicate the mean degrees of the networks.

13.5 Avalanches as branching processes

Immediately before complex networks, so-called *self-organized criticality* was the hottest topic in non-equilibrium statistical mechanics.[11] Most of the numerous researchers who studied self-organized criticality in the 1980s—1990s, later switched over to complex networks. This partially explains why many concepts and models from self-organized criticality have been extensively applied to networks. A simple example illustrates this important notion [159]. Let the nodes of an infinite lattice be occupied or empty, so that we have a set of clusters on a lattice like those in percolation problems. Consider the following process. In parallel,

[11] See the fascinating popular science book of Per Bak, a pioneer in this research field [13].

(i) infinitely slowly, fill new and new uniformly randomly selected nodes, and

(ii) also infinitely slowly, choose at random nodes and empty them together with the clusters to which these nodes belong, see Fig. 13.5.

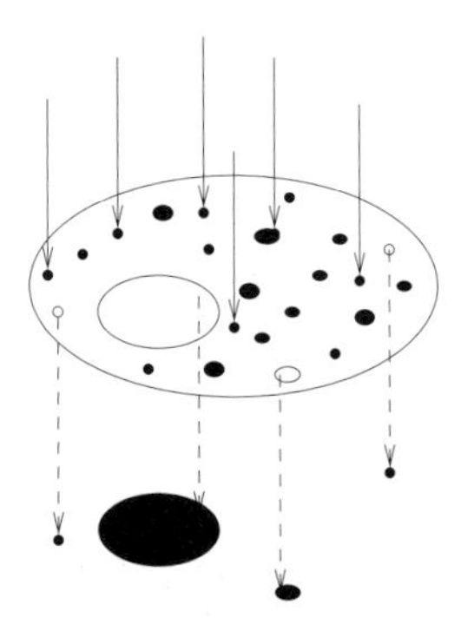

Fig. 13.5 A forest-fire model demonstrating self-organized criticality. The solid arrows show the slow filling of randomly selected nodes (growing trees). The dashed arrows the parallel infinitely slow process: random selection of nodes and elimination clusters (forests) to which these nodes belong (forest fires).

If we start from an empty lattice, the concentration of occupied nodes will slowly grow until it reaches some value. The maximal size of clusters will grow but will never approach infinity. The point is that a giant connected component cannot emerge in this system since process (ii) efficiently eliminates any cluster containing a finite fraction of nodes. Therefore the combination of slow processes (i) and (ii) drives a system exactly to the point of the emergence of a giant component, that is to a critical state. The supposed infinite slowness of the processes is crucially important. If we added nodes and removed clusters at a finite rate, the system would stay away from criticality. Remarkably, the system enters this state by itself, without fitting any control parameter, which explains the term 'self-organized criticality'. We can interpret this demonstrative example as a forest-fire model. In this interpretation, in parallel,

(i) new trees grow up in random places and form forests (clusters—connected components), and

(ii) random fires burn forests.

We can also treat each removed cluster as an avalanche initiated by the removal of a single node. It is easy to estimate the size distribution of removed clusters (avalanches or burned forests). Let us suppose for the sake of simplicity that our lattice is high-dimensional. As a first approximation, it is natural to assume that the statistics of clusters in this state is similar to that at a percolation threshold in high-dimensional lattices. Then we can use the fact that for percolation problems in this situation, the size distribution of clusters to which uniformly randomly chosen nodes belong is decreased as a power law, $s^{-3/2}$. This is the size distribution of avalanches in our model. In general, the exponent of the size distribution of avalanches in various high-dimensional self-organized criticality systems and models is exactly $3/2$.

Generally, the notion 'self-organized criticality' implies self-organization of an open non-equilibrium cooperative system into a critical state.[12] This state features power-law decaying correlations, power-law distributed avalanches, and other critical phenomena. Note the contrast: (i) in standard phase transitions, a system is driven to a critical point by setting temperature (concentration of defects, pressure, etc.) to a special critical value; (ii) in self-organized criticality, the system itself, spontaneously enters a critical state. Typically, self-organized criticality sports abrupt phenomena, and so scientists mostly study the statistics of avalanches. Among real-world self-organized criticality systems and phenomena are sandpiles, earthquakes, and many others.

After this brief insight into self-organized criticality we return to networks and place a self-organized criticality system on the top of a complex network. For the sake of brevity, we only touch upon a sandpile problem on a network [98]. Suppose that each node in a network can store grains but not more than some threshold number.[13] Let us begin slowly to add grains to randomly selected nodes.[14] The rule is that if the number of grains on a node exceeds the threshold, then some of these grains jump to randomly chosen nearest neighbours of the node.

[12] Note that in our example, the system was open. Specifically, we added nodes and removed connected components.

[13] In principle, this threshold number may depend on the number of connections of a node.

[14] As is usual in self-organized criticality, we must also incorporate a grain outflow into this scheme. For example, perpetually remove all grains from some of the nodes.

The number of toppled grains depends on a specific sandpile model. The thresholds of some of these neighbours may be exceeded, and their grains will also topple and so on. Thus the first toppling has a chance to trigger an avalanche. What is the size distribution of these avalanches? Since our networks are infinitely dimensional, we guess that this distribution is of the same form as in our demonstrative forest-fire example, namely $s^{-3/2}$. It turns out that this is indeed the case if the degree distribution of a network is a rapidly decreasing function. In the case of heavy-tailed degree distributions, the problem was resolved for infinite uncorrelated networks [98]. In these networks, thanks to their local tree-like organization, the structure of avalanches is particularly simple. Clearly, the avalanches on a network of this kind must be subgraphs of the mother network, that is trees, see Fig. 13.6. Similarly to what we observed in percolation problems, these trees–subgraphs have different degree distributions, correlations, and even dimensionality than their mother networks. Nonetheless, the similarity is not complete. In networks with heavy-tailed degree distributions, the structure and statistics of these sandpile avalanches was found to differ markedly from those of connected components in percolation problems. Furthermore, the statistics of avalanches depends significantly on a specific sandpile model. The qualitative result, valid for a wide range of self-organized criticality models, was that, counterintuitively, the presence of hubs in a network makes the size distributions of avalanches more rapidly decaying. One may say that hubs hamper large avalanches.[15] For scale-free networks with a degree distribution decaying as $q^{-\gamma}$, the size distribution of avalanches becomes exponential as exponent γ approaches 2. In this limiting situation, hubs eliminate self-organized criticality.

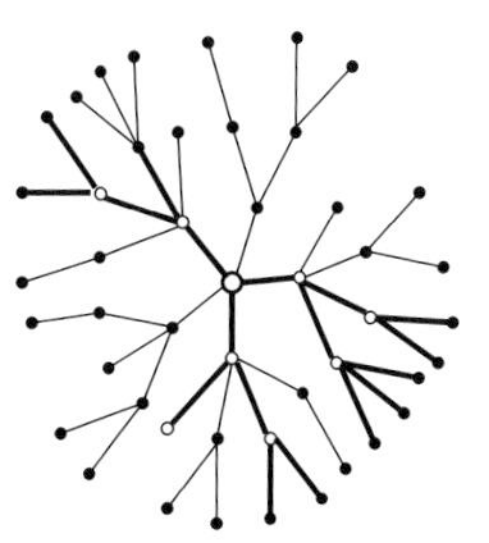

Fig. 13.6 A sandpile avalanche on a tree-like network. The large dot indicates a node which started the avalanche. Grains on white nodes were toppled, and the toppled grains jumped along the bold links to the neighbouring nodes inducing further topplings.

[15] Recall that the degree distribution of the nearest neighbour of a node is $qP(q)/\langle q \rangle$, so hubs (nodes with many connections) participate in avalanches with high probability and strongly influence their statistics.

Optimization

14

The architectures of real-world networks have one major thing in common. All of these complex architectures are optimal or nearly optimal for the function of these networks. Furthermore, the optimality of network design can be regarded as a driving force of network evolution, its engine. In this lecture we discuss how a universal requirement for optimality leads to the complex structural organization of a network.

14.1 Critique of preferential attachment

We have described the preferential attachment mechanism as a strict and convenient way for generation of heavy-tailed distributions. This explanation of scale-free distributions is particularly popular among physicists. Theoretical physicists implement preferential attachment in their constructions because it allows easy analytical treatment. In addition, preferential attachment enables researchers to rapidly generate extremely large networks for numerical simulations. Nonetheless, despite all these advantages, preferential attachment has a weak point. No satisfactory explanation of this mechanism has been proposed in numerous situations. In particular, there are no convincing quantitative arguments for specific linear forms of a preference function, necessary for scale-free distributions. Thus, preferential attachment explains power laws, but there is no good explanation for preferential attachment. A long-running criticism of the preferential attachment mechanism actually began before the complex networks boom, in the 1950s. During the 1950s—1960s, Benoît Mandelbrot, an outstanding applied mathematician who invented the fractality concept, and Herbert A. Simon, one of the proponents of what is now called the preferential attachment concept, exchanged a series of rather scaring comments on the nature of power laws. Mandelbrot repeatedly criticized the preferential attachment concept as too mechanistic and lacking in explanatory power. As an alternative to preferential attachment, he proposed consideration of optimization mechanisms [121].

In 2002, a group of network engineers unleashed another wave of criticism of preferential attachment concepts [182].[1] Their criticism was primarily aimed at a wide range of models of the organization and function of the Internet, proposed by physicists. The computer scientists stressed that these models 'are only evocative; they are not explanatory'. In their definition, an evocative model 'can reproduce the phenomenon of interest but does not necessarily capture and incorporate the true under-

[1] See also our discussion of self-similar traffic in Lecture 12.

lying cause'. On the other hand, an explanatory model 'also captures the causal mechanisms (why and how, in addition to what)'. These researchers believed that only optimization driven evolution could explain the architecture of the Internet in a natural and convincing way.

Unfortunately, this criticism from Mandelbrot and the computer scientists itself has a weak point. The problem is that optimization based models are typically far more difficult to deal with than models implementing preferential attachment. Optimization driven evolution is far less studied than preferential attachment. Mandelbrot showed very schematically how a requirement for optimality can result in power-law distributions. His work was not directly applied to networks. As for the previously mentioned group of network engineers, they did not propose 'explanatory' optimization models of networks. In principle, one can formulate plenty of models of this kind. Unfortunately, they are hardly treatable analytically or numerically. Numerical simulations of optimization driven network evolution are so time consuming that only very small networks can be generated. Their sizes (typically, a few hundred nodes) are not sufficient to arrive at solid conclusions about the structures of these networks and of their bigger counterparts. In particular, it is impossible to distinguish scale-free and non-scale-free small networks. This is why only the simplest optimization based network models were explored. In the next sections, we touch upon a few of these networks.

14.2 Optimized trade-offs

Optimization based models of networks are usually organized in the following way. New connections or rewirings must optimize some combination of network characteristics. That is, a new connection is made to the 'best place' in a network that provides the highest or smallest value of some function of characteristics of the network after this reconstruction. This function to optimize is called a *cost function*. For example, the cost function may be a linear function of the mean internode distance and other global characteristics of a network. Minimization of this function leads to more compact architectures. In another optimization scheme, the variables of a cost function are the characteristics of a node to which we link. For example, a variable can be the shortest path distance between these node, and some other one.

In 2002, three computer scientists, Fabrikant, Koutsoupias, and Papadimitriou, proposed a remarkably simple optimization driven model of a growing tree [89]. The absence of loops allowed a strict analytical treatment of the model. The growth starts from a single node (root with label 1) placed at some point of a restricted two-dimensional area. Nodes are added one by one. Each new node is placed within this area and attached to a selected existing node, which results in a tree. We can characterise completely the position of each node i in this tree by the set of distances: the shortest-path network distance $\ell_{1,i}$ between node

i and the root ('the depth' of node i) and the Euclidean distances $d_{i,j}$ between node i and other nodes, $j \neq i$, in the network, see Fig. 14.1. At each time step,

(i) place a new node t at a random point of the area and

(ii) attach it to that node i which gives the minimum cost function $\ell_{1,i} + \alpha(t)d_{i,t}$.

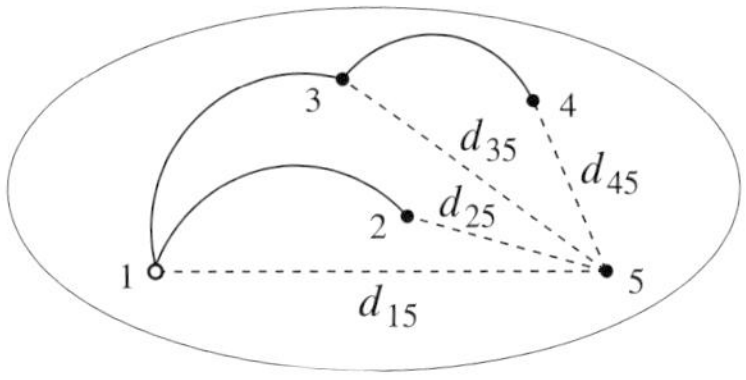

Here the coefficient $\alpha(t)$ is a given function of the network size t. This model is based on 'optimized trade-offs' between two conflicting objectives: a network objective versus a geographic one. The network objective is 'connect to the root (the centre of a network) and make the entire network more compact'. The geographic objective is 'connect to the geographically closest node', which usually means the cheapest connection. If $\alpha = 0$, then all attachments are to the root, and the network is a star, which is the most compact tree. If α is large, the attachments are mostly to the geographically closest nodes. The structure of the resulting network depends significantly on a given $\alpha(t)$. The authors of this model found that if $\alpha(t)$ is a power law, then the resulting degree distribution has a power-law part. This finding immediately made the model very popular, mostly among computer scientists. The hope was that this idea might explain scale-free of the Internet and other real networks. Unfortunately, this hope was short-lived. Already in 2003, it had been shown that the power-law region of the resulting degree distribution is narrow if the network is large [24].[2] In that sense, the Fabrikant–Koutsoupias–Papadimitriou model does not produce a really scale-free tree in the large-size limit. Only sufficiently small networks grown in this way can be perceived as scale-free.

1 2 3 4 5
d_{15} d_{25} d_{35} d_{45}

Fig. 14.1 The Fabrikant–Koutsoupias–Papadimitriou network after the first four nodes are placed within a restricted two-dimensional area. The arks show the links. d_{15}, d_{25}, d_{35}, and d_{45} are the geographic distances between the new (fifth) node and others. Find the minimum number out of αd_{15}, $1+\alpha d_{25}$, $1+\alpha d_{35}$, and $2+\alpha d_{45}$, and attach the fifth node to the corresponding node.

[2] More precisely, depending on a form of the function $\alpha(t)$, either the power-law region is quite narrow or the great majority of vertices are leaves.

14.3 The power of choice

In optimization algorithms which we discussed above, we selected the optimal node for attachment among all nodes in a network. This leads to a heavy computational task. In practice, this optimization is impossible even for moderately large networks. We can, however, do partial optimization. Namely, choose at random a number of nodes and find the optimal one among them. If this number is much smaller than the network size, then we facilitate our work dramatically. On the other hand, this optimization is only partial, and its results may be far from those of complete optimization.

Raissa D'Souza, Pavel Krapivsky, and Cristopher Moore considered a simple growing network which illustrates the essence of these algorithms [83]. This network is a growing tree, where each new node finds a node to be attached to in the following way. Select uniformly at random $k > 1$ nodes and choose the node with the highest degree among them.[3] Importantly, the number k of choices in this model must stay constant as the network grows. If the number of choices is infinite, then all attachments are to the root, which gives a star graph. In the case of

[3] The reader can see that if $k = 1$, then this network is a random recursive tree.

a finite k, the resulting degree distribution has an exponential cut-off at degrees of the order of k, which severely restricts a region, where a slower decay can be observed. For example, if $k = 2$, then this network has practically an exponential degree distribution. When k is sufficiently large, there is a range of degrees where these authors found a power-law decaying distribution. However, for reasonable numbers, say $k = 20$ or 50, this region is so narrow that it is hard to reliably observe the power law and to measure its exponent. Theoretically, the degree distribution is $P(q) \sim 1/q$ in the range $1 \ll q \ll k$. The reader can see that, indeed, the resulting architecture significantly differs from a star graph.

This choice-driven process somewhat resembles preferential attachment since highly connected nodes are preferentially selected in both of these algorithms. Nonetheless, these two mechanisms produce contrasting degree distributions with different power laws and different cut-offs. In particular, the preferential attachment mechanism cannot provide the $P(q) \sim 1/q$ degree distribution.

In 2009, Dimitris Achlioptas, Raissa D'Souza, and Joel Spencer exploited the power of choice in another problem [2]. They studied the emergence of a giant connected component in networks. To explain their problem, let us return to classical random graphs, particularly to the $G_{N,p}$ (Gilbert) model. In this notation, N is the number of nodes in the random graph, and p is the probability that two nodes in the graph are interlinked. We showed that this equilibrium random graph has a giant connected component when the mean degree of a node, $\langle q \rangle = pN$, exceeds 1, see Fig. 2.5.[4] In other words, to obtain a giant connected component, each pair of nodes must be interlinked with a probability p exceeding $1/N$. There is another view of this network. Let us start with a large set of isolated nodes. Choose uniformly at random a pair of nodes and connect them by a link. Then choose and connect another pair of nodes and so on. That is, the number of links grows, while the number of nodes is fixed. A giant connected component will emerge when the number of links reaches $N/2$, which exactly corresponds to $p = 1/N$. Achlioptas and coauthors used a similar network-formation process in their construction.

[4] Recall that the notion of a giant connected component is meaningful for infinite networks, so N tends to infinity.

Their network is also built starting with N isolating nodes, by adding links one by one. Each new link is created in the following way, see Fig. 14.2.

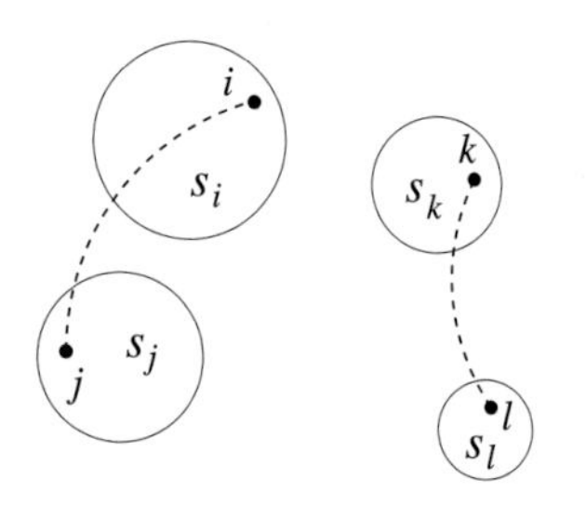

Fig. 14.2 The choice of nodes to connect by a link in the scheme of Achlioptas, D'Souza, and Spencer. Of these two pairs of nodes, the right one must be chosen since the product of the sizes of connected components to which its nodes belong, $s_k s_l$, is smaller than $s_i s_j$.

(i) Pick uniformly at random two pairs of nodes, say, i, j and k, l. Each node in a network belongs to some connected components. For each of the two pairs compute the products of the sizes of connected components to which these nodes belong, $s_i s_j$ and $s_k s_l$.

(ii) Create a link between that pair of nodes which has the smallest product.[5]

[5] In another version of this process, the sums $s_i + s_j$ and $s_k + s_l$ are compared instead of the products, which produce similar phenomena.

These rules lead to quite different statistics of connected components than in classical random graphs. We can treat the forming of this network as a specific aggregation process in which selected pairs of components merge into single ones. Importantly, the rules hamper the creation of large connected components and ultimately delay the emergence of a giant connected component. As a result, a giant connected component emerges at a higher concentration of connections than in classical random graphs. Moreover, according to Achlioptas and coauthors, this component emerges abruptly, as a sudden jump, so that this phenomenon was named *explosive percolation*, see Fig. 14.3. At the end of 2009, it is still unclear if this jump is supplemented by a singularity as for a hybrid transition (recall the emergence of a k-core). We can increase the number of choices in this scheme. For example, instead of comparing two pairs of nodes, we can compare three or more pairs. Then a giant component will emerge at a higher concentration of connections and a critical jump will be bigger.

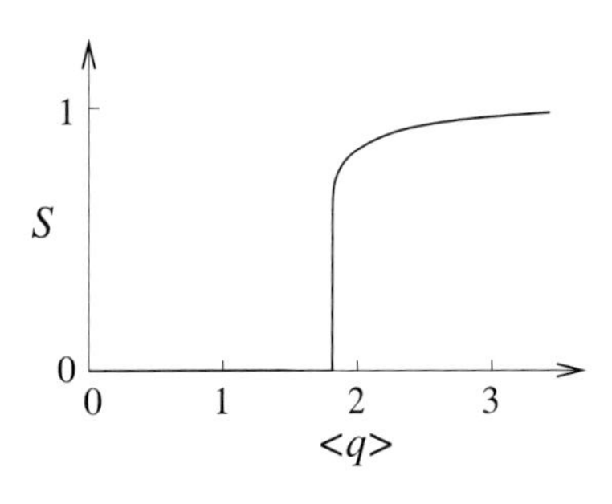

Fig. 14.3 Schematic plot of the dependence of the relative size of a giant connected component on the mean degree $\langle q \rangle$ in this construction according to Achlioptas and coauthors [2].

15 Outlook

Three major milestones mark the history of the exploration of networks: Leonhard Euler's work (1735), the introduction of random graphs (1950s), and the launch of the large-scale study of complex networks (the end of the 1990s, although isolated works on the topic appeared earlier). Each of the new milestones was related to a wider range of real-world network systems and to a wider range of sciences involved. The boom in networks, starting in the 1990s, has already involved and affected practically all natural and human sciences. Surely, the emergence of the Internet and WWW have played a pivotal role in the latest advance. So, the science of networks in its currents state is an ultimately multidisciplinary research field. We can hardly expect such dramatic breakthroughs in the near future, but the incredible rate of progress in the understanding of networks apparently does not slow down. Below we list a few particularly hot and prospective topics in complex networks.

(1) Networks with rich community structures, community detection.
(2) Complex hypergraphs and multi-partite networks.
(3) Loops and motifs in networks.
(4) Flows in complex networks.
(5) Optimization driven design of networks.
(6) Coevolving networks and interacting systems.
(7) Dynamical systems placed on networks.
(8) Searching, retrieving, and indexing information from complex networked environments.

In the long run, the task is to translate the general understanding of networks into working methods and strategies. The ultimate aim is to control, manipulate, and use the structure and function of real-world networks. Despite recent spectacular advances, we are still a long way from this challenging goal.

Further reading

Here is a short, incomplete list of popular science books, reference books, and comprehensive reviews on complex networks.

(1) Watts, D. J. (1999). *Small Worlds: The Dynamics of Networks between Order and Randomness*. Princeton University Press, Princeton.

(2) Barabási, A.-L. (2002). *Linked: The New Science of Networks*. Perseus, Cambridge MA.

(3) Dorogovtsev, S. N. and Mendes, J. F. F. (2003). *Evolution of Networks: From the Biological Nets to the Internet and WWW*. Oxford University Press, Oxford.

(4) Pastor-Satorras, R. and Vespignani, A. (2004). *Evolution and Structure of the Internet: A Statistical Physics Approach*. Cambridge University Press, Cambridge.

(5) Watts, D. J. (2003). *Six Degrees: The Science of a Connected Age*. Norton, New York.

(6) Barrat, A., Barthélemy, M., and Vespignani, A. (2008). *Dynamical Processes on Complex Networks*. Cambridge University Press, Cambridge.

(7) Caldarelli, G. (2007). *Scale-Free Networks: Complex Webs in Nature and Technology*. (Oxford Finance) Oxford University Press, Oxford.

(8) Albert, R. and Barabási, A.-L. (2002). Statistical mechanics of complex networks. *Rev. Mod. Phys.* **47**, 74.

(9) Dorogovtsev, S. N. and Mendes, J. F. F. (2002). Evolution of networks. *Adv. Phys.* **51**, 1079.

(10) Newman, M. E. J. (2003). The structure and function of complex networks. *SIAM Review* **45**, 167.

(11) Boccaletti, S., Latora, V., Moreno, Y., Chavez, M., and Hwang, D.-U. (2006). Complex networks: structure and dynamics. *Phys. Rep.* **424**, 175.

(12) Dorogovtsev, S. N., Goltsev, A. V., and Mendes, J. F. F. (2008). Critical phenomena in complex networks. *Rev. Mod. Phys.* **80**, 1275; arXiv.org/0705.0010.

(13) Arenas, A., Diaz-Guilera, A., Kurths, J., Moreno, Y., and Zhou, C. (2008). Synchronization in complex networks. *Phys. Rep.* **469**, 93.

(14) Costa, L. F., Rodrigues, F. A., Travieso, G., and Villas Boas, P. R. (2007). Characterization of complex networks: A survey of measurements. *Adv. Phys.* **56**, 167.

(15) Castellano, C., Fortunato, S., and Loreto, V. (2009). Statistical physics of social dynamics. *Rev. Mod. Phys.* **81**, 591.

References

[1] Abramowicz, H., *et al.* (2009). *Future Internet: The Cross-ETP Vision Document.* (D. Papadimitriou, ed.) http://www.future-internet.eu/fileadmin/documents/reports/Cross-ETPs_FI_Vision_Document_v1_0.pdf.

[2] Achlioptas, D., D'Souza, R. M., and Spencer, J. (2009). Explosive percolation in random networks. *Science* **1453**, 323.

[3] Adler, J. (1991). Bootstrap percolation. *Physica A* **171**, 453.

[4] Albert, R. and Barabási, A.-L. (2002). Statistical mechanics of complex networks. *Rev. Mod. Phys.* **47**, 74.

[5] Albert, R., Jeong, H., and Barabási, A.-L. (1999). Diameter of the world-wide web. *Nature* **401**, 130.

[6] Albert, R., Jeong, H., and Barabási, A.-L. (2000). Attack and error tolerance of complex networks. *Nature* **406**, 378.

[7] Aleksiejuk, A., Holyst, J. A., and Stauffer, D. (2002). Ferromagnetic phase transition in Barabási-Albert networks. *Physica A* **310**, 260.

[8] Almaas, E., Kovács, B., Vicsek, T., Oltvai, Z. N., and Barabási, A.-L. (2004). Global organization of metabolic fluxes in the bacterium, *Escherichia coli*. *Nature* **427**, 839.

[9] Alvarez-Hamelin, J. I., Dall´Asta, L., Barrat, A., and Vespignani, A. (2006). k-core decomposition: a tool for the visualization of large scale networks. *Advances in Neural Information Processing Systems (Canada)* **18**, 41.

[10] Amaral, L. A. N., Scala, A., Barthélémy, M., and Stanley H. E. (2000). Classes of small-world networks. *PNAS* **97**, 11149.

[11] Arenas, A., Díaz-Guilera, A., and Guimerà, R. (2001). Communication in networks with hierarchical branching. *Phys. Rev. Lett.* **86**, 3196.

[12] Arenas, A., Diaz-Guilera, A., Kurths, J., Moreno, Y., and Zhou, C. (2008). Synchronization in complex networks. *Phys. Rep.* **469**, 93.

[13] Bak, P. (1996). *How Nature Works: The Science of Self-Organized Criticality*. New York, Copernicus.

[14] Barabási, A.-L. and Albert, R. (1999). Emergence of scaling in random networks. *Science* **286**, 509.

[15] Barabási, A.-L. and Oltvai, Z. N. (2004). Network biology: understanding the cell's functional organization. *Nature Reviews Genetics* **5**, 101.

[16] Barabási, A.-L., Ravasz, E., and Vicsek, T. (2001). Deterministic scale-free networks. *Physica A* **299**, 559.

[17] Barahona, M. and Pecora, L. (2002). Synchronization in small-world systems. *Phys. Rev. Lett.* **89**, 054191.

[18] Barrat, A., Barthélémy, M., and Vespignani, A. (2004). Weighted evolv-

ing networks: coupling topology and weights dynamics. *Phys. Rev. Lett.* **92**, 228701.

[19] Barrat, A., Barthélémy, M., Pastor-Satorras, R., and Vespignani, A. (2004). The architecture of complex weighted networks. *PNAS* **101**, 3747.

[20] Barrat, A. and Pastor-Satorras, R. (2005). Rate equation approach for correlations in growing network models. *Phys. Rev. E* **71**, 036127.

[21] Baskerville, K., Grassberger, P., and Paczuski, M. (2007). Graph animals, subgraph sampling, and motif search in large networks. *Phys. Rev. E* **76**, 036107.

[22] Bender, E. A. and Canfield, E. R. (1978). The asymptotic number of labelled graphs with given degree sequences. *J. Combinatorial Theor. A* **24**, 296.

[23] Bénichou, O. and Voituriez, R. (2007). Comment on 'Localization transition of biased random walks on random networks'. *Phys. Rev. Lett.* **99**, 209801.

[24] Berger, N., Bollobás, B., Borgs, C., Chayes, J., and Riordan, O. (2003). Degree distribution of the FKP network model. *Lecture Notes in Computer Science* **2719**, 725.

[25] Berners-Lee, T. (1999). *Weaving the Web: The Past, Present and Future of the World Wide Web by Its Inventor*. Harper Collins Publishers, London.

[26] Bianconi, G. (2002). Mean field solution of the Ising model on a Barabási-Albert network. *Phys. Lett. A* **303**, 166.

[27] Bianconi, G. (2005). Emergence of weight-topology correlations in complex scale-free networks. *Europhys. Lett.* **71**, 1029.

[28] Bianconi, G. and Barabási, A.-L. (2001a). Competition and multiscaling in evolving networks. *Europhys. Lett.* **54**, 439.

[29] Bianconi, G. and Barabási, A.-L. (2001b). Bose–Einstein condensation in complex networks. *Phys. Rev. Lett.* **86**, 5632.

[30] Bianconi, G. and Capocci, A. (2003). Number of loops of size h in growing scale-free networks. *Phys. Rev. Lett.* **90**, 078701.

[31] Bianconi, G., Gulbahce, N., and Motter, A. E. (2008). Local structure of directed networks. *Phys. Rev. Lett.* **100**, 118701.

[32] Bianconi, G. and Marsili, M. (2005). Loops of any size and Hamilton cycles in random scale-free networks. *J. Stat. Mech.* P06005.

[33] Bianconi, G. and Marsili, M. (2006). Emergence of large cliques in random scale-free networks. *Europhys. Lett.* **74**, 740.

[34] Boguñá, M. and Krioukov, D. (2009). Navigating ultrasmall worlds in ultrashort time. *Phys. Rev. Lett.* **102**, 058701.

[35] Boguñá, M., Krioukov, D., and claffy, kc. (2009). Navigability of complex networks. *Nature Physics* **5**, 74.

[36] Boguñá, M. and Pastor-Satorras, R. (2002) Epidemic spreading in correlated complex networks. *Phys. Rev. E* **66**, 047104.

[37] Boguñá, M., Pastor-Satorras, R., and Vespignani, A. (2003) Epidemic spreading in complex networks with degree correlations. *Lecture Notes in Physics* **625**, 127.

[38] Bollobás, B. (1980). A probabilistic proof of an asymptotic formula for the number of labelled random graphs. *Eur. J. Combinatorics* **1**, 311.

[39] Bollobás, B. (1984). The evolution of sparse graphs. In *Graph Theory and Combinatorics: Proc. Cambridge Combinatorial Conf. in honour of Paul Erdős* (B. Bollobás, ed.). Academic Press, New York, p. 35.

[40] Bollobás, B. and Riordan, O. M. (2002). Mathematical results on scale-free random graphs. In *Handbook of Graphs and Networks: From the Genome to the Internet* (S. Bornholdt and H. G. Schuster, eds.). Wiley-VCH, Berlin, p. 1.

[41] Bollobás, B. and Riordan, O. M. (2004). The diameter of a scale-free random graph. *Combinatorica* **24**, 5.

[42] Bornholdt, S. and Ebel, H. (2001). World-Wide Web scaling exponent from Simon's 1955 model. *Phys. Rev. E* **64**, 035104.

[43] Brin, S. and Page, L. (1998). The anatomy of a large-scale hypertextual web search engine. In *Proc. of the Seventh Int. World Wide Web Conf.* p. 107.

[44] Broder, A., Kumar, R., Maghoul, F., Raghavan, P., Rajagopalan, S., Stata, R., Tomkins, A., and Wiener, J. (2000). Graph structure in the web. *Comput. Netw.* **33**, 309.

[45] Burda, Z., Correia, J. D., and Krzywicki, A. (2001). Statistical ensemble of scale-free random graphs. *Phys. Rev. E* **64**, 046118.

[46] Burda, Z., Jurkiewicz, J., and Krzywicki, A. (2004). Network transitivity and matrix models. *Phys. Rev. E* **69**, 026106.

[47] Burda, Z. and Krzywicki, A. (2003). Uncorrelated random networks. *Phys. Rev. E* **67**, 046118.

[48] Caldarelli, G., Capocci, A., De Los Rios, P., and Muñoz, M. A. (2002). Scale-free networks from varying vertex intrinsic fitness. *Phys. Rev. Lett.* **89**, 258702.

[49] Callaway, D. S., Hopcroft, J. E., Kleinberg, J. M., Newman, M. E. J., and Strogatz, S. H. (2001). Are randomly grown graphs really random? *Phys. Rev. E* **64**, 041902.

[50] Callaway, D. S., Newman, M. E. J., Strogatz, S. H., and Watts, D. J. (2000). Network robustness and fragility: Percolation on random graphs. *Phys. Rev. Lett.* **85**, 5468.

[51] Carmi, S., Carter, S., Sun, J., and ben-Avraham, D. (2009). Asymptotic behavior of the Kleinberg model. *Phys. Rev. Lett.* **102**, 238702.

[52] Cartozo, C. C. and De Los Rios, P. (2009). Extended navigability of small world networks: exact results and new insights. *Phys. Rev. Lett.* **102**, 238703.

[53] Cerf, V. (2001). A brief history of the Internet and related networks. http://www.isoc.org/internet/history/cerf.shtml.

[54] Chalupa, J., Leath, P. L., and Reich, G. R. (1979). Bootstrap percolation on a Bethe lattice. *J. Phys. C* **12**, L31.

[55] Chung, F. and Lu, L. (2002). Connected components in random graphs with given degree sequences. *Ann. Combin.* **6**, 125.

[56] Clauset, A. and Moore, C. (2005). Accuracy and scaling phenomena in Internet mapping. *Phys. Rev. Lett.* **94**, 018701.

[57] Cohen, R., ben-Avraham, D., and Havlin S. (2002). Percolation critical exponents in scale-free networks. *Phys. Rev. E* **66**, 036113.

[58] Cohen, R., Erez, K., ben-Avraham, D., and Havlin, S. (2000). Resilience of the Internet to random breakdowns. *Phys. Rev. Lett.* **85**, 4625.

[59] Cohen, R., Erez, K., ben-Avraham, D., and Havlin, S. (2001). Breakdown of the Internet under intentional attack. *Phys. Rev. Lett.* **86**, 3682.

[60] Cohen, R. and Havlin, S. (2003). Ultra small world in scale-free networks. *Phys. Rev. Lett.* **90**, 058701.

[61] Cohen, R., Havlin, S., and ben-Avraham, D. (2003a). Structural properties of scale free networks. In *Handbook of Graphs and Networks: From the Genome to the Internet* (S. Bornholdt and H. G. Schuster, eds.). Wiley-VCH, Berlin, p. 85.

[62] Cohen, R., Havlin, S., and ben-Avraham, D. (2003b). Efficient immunization strategies for computer networks and populations. *Phys. Rev. Lett.* **91**, 247901.

[63] Colizza V., Flammini A., Serrano, M. A., and Vespignani A. (2006). Detecting rich-club ordering in complex networks. *Nature Physics* **2**, 110.

[64] Colizza, V. and Vespignani, A. (2007). Invasion threshold in heterogeneous metapopulation networks. *Phys. Rev. Lett.* **99**, 148701.

[65] Connolly, D. (2000). A little history of the World Wide Web. http://www.w3.org/History.html.

[66] Costa, L. da F., Rodrigues, F. A., Travieso, G., and Boas, P. R. Villas. (2007). Characterization of complex networks: A survey of measurements. *Adv. Phys.* **56**, 167.

[67] Dall, J. and Christensen, M. (2002). Random geometric graphs. *Phys. Rev. E* **66**, 016121.

[68] de Menezes, M. A. and Barabási, A.-L. (2004). Fluctuations in network dynamics. *Phys. Rev. Lett.* **92**, 028701.

[69] Derényi, I., Palla G., and Vicsek, T. (2005). Clique percolation in random networks. *Phys. Rev. Lett.* **94**, 160202.

[70] Dodds, P. S., Muhamad, R., and Watts, D. J. (2003). An experimental study of search in global social networks. *Science* **301**, 827.

[71] Donato, D., Laura, L., Leonardi, S., and Millozzi, S. (2004). Large scale properties of the Webgraph. *Eur. Phys. J. B* **38**, 239.

[72] Donetti, L., Hurtado, P. I., and Muñoz, M. A. (2005). Entangled networks, synchronization, and optimal network topology. *Phys. Rev. Lett.* **95**, 188701.

[73] Dorogovtsev, S. N., Ferreira, A. L., Goltsev, A. V., and Mendes, J. F. F. (2009). Zero Pearson coefficient for strongly correlated growing trees. arXiv.org/0911.4285.

[74] Dorogovtsev, S. N., Goltsev, A. V., and Mendes, J. F. F. (2002a). Pseudofractal scale-free web. *Phys. Rev. E* **65**, 066122.

[75] Dorogovtsev, S. N., Goltsev, A. V., and Mendes, J. F. F. (2002b). Ising model on networks with an arbitrary distribution of connections. *Phys. Rev. E* **66**, 016104.

[76] Dorogovtsev, S. N., Goltsev, A. V., and Mendes, J. F. F. (2006). k-core organization of complex networks. *Phys. Rev. Lett.* **96**, 040601.

[77] Dorogovtsev, S. N., Goltsev, A. V., and Mendes, J. F. F. (2008). Critical phenomena in complex networks. *Rev. Mod. Phys.* **80**, 1275; arXiv.org/0705.0010.

[78] Dorogovtsev, S. N. and Mendes, J. F. F. (2001). Effect of the accelerated growth of communications networks on their structure. *Phys. Rev. E* **63**, 025101.

[79] Dorogovtsev, S. N. and Mendes, J. F. F. (2002). Evolution of networks. *Adv. Phys.* **51**, 1079.

[80] Dorogovtsev, S. N. and Mendes, J. F. F. (2003). *Evolution of Networks: From the Biological Nets to the Internet and WWW*. Oxford University Press, Oxford.

[81] Dorogovtsev, S. N., Mendes, J. F. F., and Samukhin, A. N. (2000). Structure of growing networks with preferential linking. *Phys. Rev. Lett.* **85**, 4633.

[82] Dorogovtsev, S. N., Mendes, J. F. F., and Samukhin, A. N. (2001). Anomalous percolation properties of growing networks. *Phys. Rev. E* **64**, 066110.

[83] D'Souza, R. M., Krapivsky, P. L., and Moore, C. (2007). The power of choice in network growth. *Eur. Phys. J. B* **59**, 535.

[84] Echenique, P., Gómez-Gardeñes, J., and Moreno, Y. (2005). Dynamics of jamming transitions in complex networks. *Europhys. Lett.* **71**, 325.

[85] Eckmann, J.-P. and Moses, E. (2002). Curvature of co-links uncovers hidden thematic layers in the World Wide Web. *PNAS* **99**, 5825.

[86] Emmerling, M., Dauner, M., Ponti, A., Fiaux, J., Hochuli, M., Szyperski, T., Wuthrich, K., Bailey, J. E., Sauer, U. (2002). Metabolic flux responses to pyruvate kinase knockout in Escherichia coli. *J. Bacteriology* **184**, 152.

[87] Erdős, P. and Rényi, A. (1959). On random graphs. *Publ. Math. Debrecen* **6**, 290.

[88] Erdős, P. and Rényi, A. (1960). On the evolution of random graphs. *Publ. Math. Inst. Hung. Acad. Sci.* **5**, 17.

[89] Fabrikant, A., Koutsoupias, E., and Papadimitriou, C. H. (2002). Heuristically optimized trade-offs: A new paradigm for power laws in the Internet. *Lecture Notes in Computer Science* **2380**, 110.

[90] Fabrikant, A., Luthra, A., Maneva, E., Papadimitriou, C. H., and Shenker, S. (2003). On a network creation game. In *Proceedings. of the 22nd ACM Symposium on Principles of Distributed Computing (PODC 2003)* p. 347.

[91] Faloutsos, M., Faloutsos, P., and Faloutsos, C. (1999). On power-law relationships of the Internet topology. *Comput. Commun. Rev.* **29**, 251.

[92] Feng, Q., Su, C., and Hu, Z. (2005). Branching structure of uniform recursive trees. *Science in China Ser. A* **48**, 769.

[93] Fortunato, S., Boguñá, M., Flammini, A., and Menczer, F. (2008). Approximating PageRank from in-degree. *Lecture Notes In Computer Science* **4936**, 59.

[94] Fortunato, S. and Castellano, C. (2009). Community structure in graphs. In *Encyclopedia of Complexity and Systems Science* (R. A. Meyers, ed.). Springer, Berlin, p. 1141; arXiv:0712.2716.

[95] Fronczak, A. and Fronczak, P. (2009). Biased random walks on complex networks: the role of local navigation rules. *Phys. Rev. E* **80**, 016107.

[96] Goh, K.-I., Cusick, M. E., Valle, D., Childs, B., Vidal, M., and Barabási, A.-L. (2007). The human disease network. *PNAS* **104**, 8685.

[97] Goh, K.-I., Kahng, B., and Kim, D. (2001). Universal behavior of load distribution in scale-free networks. *Phys. Rev. Lett.* **87**, 278701.

[98] Goh, K.-I., Lee, D. S., Kahng, B., and Kim, D. (2003). Sandpile on scale-free networks. *Phys. Rev. Lett.* **91**, 148701.

[99] Goh, K.-I., Salvi, G., Kahng, B., and Kim, D. (2006). Skeleton and fractal scaling in complex networks. *Phys. Rev. Lett.* **96**, 018701.

[100] Granovetter, M. S. (1973). The strength of weak ties. *American Journal of Sociology* **78**, 1360.

[101] Granovetter, M. (1983). The strength of weak ties: A network theory revisited. *Sociological Theory* **1**, 201.

[102] Hartmann, A. K. and Weigt, M. (2005). *Phase Transitions in Combinatorial Optimization Problems: Basics, Algorithms and Statistical Mechanics.* Wiley-VCH, Berlin.

[103] Hong, H., Kim B. J., Choi, M. Y., and Park, H. (2004). Factors that predict better synchronizability on complex networks. *Phys. Rev. E* **69**, 067105.

[104] Ichinomiya, T. (2004). Frequency synchronization in a random oscillator network. *Phys. Rev. E* **70**, 026116.

[105] Jeong, H., Mason, S.P., Barabási, A.-L., and Oltvai, Z.N. (2001). Lethality and centrality in protein networks. *Nature* **411**, 41.

[106] Jeong, H., Tombor, B., Albert, R., Oltvai, Z.N., and Barabási, A.-L. (2000). The large-scale organization of metabolic networks. *Nature* **407**, 651.

[107] Jonsson, T. and Stefánsson, S. Ö. (2008). The spectral dimension of random brushes. *J. Phys. A* **41**, 045005.

[108] Kim, D.-H. and Motter, A. E. (2007). Ensemble averageability in network spectra. *Phys. Rev. Lett.* **98**, 248701.

[109] Kim, B. J., Trusina, A., Minnhagen, P., and Sneppen, K. (2005). Self organized scale-free networks from merging and re-generation. *Eur. Phys. J. B* **43**, 369.

[110] Kleinberg, J. (2000). Navigation in a small world. *Nature* **406**, 845.

[111] Kleinberg, J. (2000). The small-world phenomenon: An algorithmic perspective. In *Proc. 32nd ACM Symposium on Theory of Computing* p. 163.

[112] Kleinberg, J. (2006). Complex networks and decentralized search algorithms. In *Proc. Int. Congress of Mathematicians (ICM) III* 1019.

[113] Kleinfeld, J. (2002). The small world problem. (Could it be a big world after all? The 'six degrees of separation' myth.). *Society*, **39**, 62.

[114] Krapivsky, P. L, Redner, S., and Leyvraz, F. (2000). Connectivity of growing random networks. *Phys. Rev. Lett.* **85**, 4629.

[115] Krapivsky, P. L. and Redner, S. (2001). Organization of growing random networks. *Phys. Rev. E* **63**, 066123.

[116] Kuramoto, Y. (1984). *Chemical Oscillations, Waves, and Turbulence.*

Springer, Berlin.

[117] Lee, E. J., Goh, K.-I., Kahng, B., and Kim, D. (2005). Robustness of the avalanche dynamics in data packet transport on scale-free networks. *Phys. Rev. E* **71**, 056108.

[118] Leland, W. E., Taqqu, M. S., Willinger, W., and Wilson, D. W. (1994). On the self-similar nature of Ethernet traffic (extended version). *IEEE/ ACM Trans. Network.* **2**, 1.

[119] Leone, M., Vázquez, A., Vespignani, A., and Zecchina, R. (2002). Ferromagnetic ordering in graphs with arbitrary degree distribution. *Eur. Phys. J. B* **28**, 191.

[120] Leskovec, J., Kleinberg, J., and Faloutsos, C. (2007). Laws of graph evolution: Densification and shrinking diameters. *ACM Transactions on Knowledge Discovery from Data (ACM TKDD)* **1** (1), article No 2.

[121] Mandelbrot, B. (1953). An informational theory of the statistical structure of languages. *Communication theory*, (W. Jackson, ed.) Betterworth, London, p. 486.

[122] Maslov, S. and Sneppen, K. (2002). Specificity and stability in topology of protein networks. *Science* **296**, 910.

[123] Maslov, S., Sneppen, K., and Alon, U. (2003). Correlation profiles and motifs in complex networks. In *Handbook of Graphs and Networks: From the Genome to the Internet* (S. Bornholdt and H. G. Schuster, eds.). Wiley-VCH, Berlin, p. 168.

[124] Meloni, S., Gomez-Gardenes, J., Latora, V., and Moreno, Y. (2008). Scaling breakdown in flow fluctuations on complex networks. *Phys. Rev. Lett.* **100**, 208701.

[125] Milgram, S. (1967). The small world problem. *Psychology Today* **2**, 60.

[126] Milo, R., Shen-Orr, S., Itzkovitz, S., Kashtan, N., Chklovskii, D., and Alon, U. (2002). Network motifs: Simple building blocks of complex networks. *Science* **298**, 824.

[127] Mitzenmacher, M. (2001). A brief history of generative models for power law and lognormal distributions. Harvard University, Computer Science Group, Technical Report TR-08-01.

[128] Molloy, M. and Reed, B. A. (1995). A critical point for random graphs with a given degree sequence. *Random Struct. Algor.* **6**, 161.

[129] Molloy, M. and Reed, B. A. (1998). The size of the giant component of a random graph with a given degree sequence. *Combin. Prob. Comp.* **7**, 295.

[130] Moreno, Y. and Pacheco, A. F. (2004). Synchronization of Kuramoto oscillators in scale-free networks. *Europhys. Lett.* **68**, 603.

[131] Motter, A. E. and Lai, Y.-C. (2002). Cascade-based attacks on complex networks. *Phys. Rev. E* **66**, 065102.

[132] Newman, M. E. J. (2000). Models of the small world. *J. Stat. Phys.* **101**, 819.

[133] Newman, M. E. J. (2002). Random graphs as models of networks. In *Handbook of Graphs and Networks: From the Genome to the Internet* (S. Bornholdt and H. G. Schuster, eds.). Wiley-VCH, Berlin, p. 35.

[134] Newman, M. E. J. (2002). Assortative mixing in networks. *Phys. Rev.*

Lett. **89**, 208701.

[135] Newman, M. E. J. (2006). Finding community structure in networks using the eigenvectors of matrices. *Phys. Rev. E* **74**, 036104.

[136] Newman, M. E. J. (2007). Component sizes in networks with arbitrary degree distributions. *Phys. Rev. E* **76**, 045101.

[137] Newman, M. E. J. (2009). Random graphs with clustering. *Phys. Rev. Lett.* **103**, 058701.

[138] Newman, M. E. J., Forrest, S., and Balthrop, J. (2002). Email networks and the spread of computer viruses. *Phys. Rev. E* **66**, 035101.

[139] Newman, M. E. J. and Girvan, M. (2004). Finding and evaluating community structure in networks. *Phys. Rev. E* **69**, 026113.

[140] Newman M. E. J., Strogatz, S. H., and Watts D. J. (2001). Random graphs with arbitrary degree distributions and their applications. *Phys. Rev. E* **64**, 026118.

[141] Noh, J. D. and Rieger, H. (2004). Random walks on complex networks. *Phys. Rev. Lett.* **92**, 118701.

[142] Nowak, M. A. and May, R. M. (1992). Evolutionary games and spatial chaos. *Nature* **359**, 826.

[143] Ohno, S. (1970). *Evolution by Gene Duplication.* Springer-Verlag, New York.

[144] Onnela, J.-P., Saramäki, J., Hyvönen, J., Szabó, G., Lazer, D., Kaski, K., Kertész, J., and Barabási, A.-L. (2007). Structure and tie strengths in mobile communication networks. *PNAS* **104**, 7332.

[145] Onnela, J.-P., Saramäki, J., Hyvönen, J., Szabó, G., de Menezes, M., Kaski, K., Barabási, A.-L., and Kertész, J. (2007). Analysis of a large-scale weighted network of one-to-one human communication. *New J. Phys.* **9**, 179.

[146] Pacheco, J. M., Traulsen, A., and Nowak, M. A. (2006). Co-evolution of strategy and structure in complex networks with dynamical linking. *Phys. Rev. Lett.* **97**, 258103.

[147] Palla, G., Derenyi, I., Farkas, I., and Vicsek, T. (2005). Uncovering the overlapping community structure of complex networks in nature and society. *Nature* **435**, 814.

[148] Pastor-Satorras, R., Vázquez, A., and Vespignani, A. (2001). Dynamical and correlation properties of the Internet. *Phys. Rev. Lett.* **87**, 258701.

[149] Pastor-Satorras, R. and Vespignani, A. (2001). Epidemic spreading in scale-free networks. *Phys. Rev. Lett.* **86**, 3200.

[150] Pastor-Satorras, R. and Vespignani, A. (2004). *Evolution and Structure of the Internet: A Statistical Physics Approach.* Cambridge University Press, Cambridge.

[151] Petermann, T. and De Los Rios. P. (2004). Exploration of scale-free networks: do we measure the real exponents? *Euro. Phys. J. B* **38**, 201.

[152] Price, D. J. de S. (1965). Networks of scientific papers. *Science* **149**, 510.

[153] Price, D. J. de S. (1976). A general theory of bibliometric and other cumulative advantage processes. *J. Amer. Soc. Inform Sci.* **27**, 292.

[154] Ravasz, E. and Barabási, A.-L. (2003). Hierarchical organization in

complex networks. *Phys. Rev. E* **67**, 026112.

[155] Ravasz, E., Somera, A. L., Mongru, D. A., Oltvai, Z. N., and Barabasi, A.-L. (2002). Hierarchical organization of modularity in metabolic networks. *Science* **297**, 1551.

[156] Rho, K., Jeong, H., and Kahng, B. (2006). Identification of lethal cluster of genes in the yeast transcription network. *Physica A* **364**, 557.

[157] Rodgers, G. J. and Bray, A. J. (1988). Density of states of a sparse random matrix. *Phys. Rev. B* **37**, 3557.

[158] Santos, F. C. and Pacheco, J. M. (2005). Scale-free networks provide a unifying framework for the emergence of cooperation. *Phys. Rev. Lett.* **95**, 098104.

[159] Schenk, K., Drossel, B., Clar, S., and Schwabl, F. (2000). Finite-size effects in the self-organized critical forest-fire model. *Eur. Phys. J. B* **15**, 177.

[160] Serrano, M. A., Krioukov, D., and Boguñá, M. (2008). Self-similarity of complex networks and hidden metric spaces. *Phys. Rev. Lett.* **100**, 078701.

[161] Shao, J., Buldyrev, S. V, Cohen, R., Kitsak, M., Havlin, S., and Stanley, H. E. (2008). Fractal boundaries of complex networks. *EPL* **84**, 48004.

[162] Simkin, M. V. and Roychowdhury, V. P. (2006). Re-inventing Willis. physics/0601192.

[163] Simon, H. A. (1955). On a class of skew distribution functions. *Biometrica* **42**, 425.

[164] Smith, R. D. (2008). The dynamics of Internet traffic: Self-similarity, self-organization, and complex phenomena. arXiv:0807.3374.

[165] Solomonoff, R. and Rapoport, A. (1951). Connectivity of random nets. *Bulletin of Mathematical Biophysics* **13**, 107.

[166] Solomonoff, R. (1952). An exact method for the computation of the connectivity of random nets. *Bulletin of Mathematical Biophysics* **14**, 153.

[167] Sood, V. and Grassberger, P. (2007). Localization transition of biased random walks on random networks. *Phys. Rev. Lett.* **99**, 098701.

[168] Sreenivasan, S., Cohen, R., López, E., Toroczkai, Z., and Stanley, H. E. (2007). Structural bottlenecks for communication in networks. *Phys. Rev. E* **75**, 036105.

[169] Stauffer, D. and Aharony, A. (1992). *Introduction to Percolation Theory*. Taylor & Francis, London.

[170] Strauss, D. (1986). On a general class of models for interaction. *SIAM Review* **28**, 513.

[171] Strogatz S. H. (2003). *SYNC: The Emerging Science of Spontaneous Order*. Hyperion, New York.

[172] Strogatz S. H. (2000). From Kuramoto to Crawford: Exploring the onset of synchronization in populations of coupled oscillators. *Physica D* **143**, 1.

[173] Szabó, G. and Fath, G. (2007). Evolutionary games on graphs. *Phys. Rep.* **446**, 97.

[174] Szymański, J. (1987). On a nonuniform random recursive tree. *Ann. Discrete. Math.* **33**, 297.

[175] Tadic, B., Thurner, S., and Rodgers G. J. (2004). Traffic on complex networks: Towards understanding global statistical properties from microscopic density fluctuations. *Phys. Rev. E* **69**, 036102.

[176] Vázquez, A., Dobrin, R., Sergi, D., Eckmann, J.-P., Oltvai, Z. N., and Barabási, A.-L. (2004). The topological relationship between the large-scale attributes and local interaction patterns of complex networks. *PNAS* **101**, 17945.

[177] Vázquez, A., Pastor-Satorras, R., and Vespignani, A. (2002). Large-scale topological and dynamical properties of Internet. *Phys. Rev. E* **65**, 066130.

[178] Valverde, S. and Solé, R. V. (2005). Logarithmic growth dynamics in software networks. *Europhys. Lett.* **72**, 858.

[179] Wang, W.-X., Wang, B.-H., Yin, C.-Y., Xie, Y.-B., and Zhou, T. (2006). Traffic dynamics based on local routing protocol on a scale-free network. *Phys. Rev. E* **73**, 026111.

[180] Watts, D. J., Dodds, P. S., and Newman, M. E. J. (2002). Identity and search in social networks. *Science* **296**, 1302.

[181] Watts, D. J. and Strogatz, S. H. (1998). Collective dynamics of small-world networks. *Nature* **393**, 440.

[182] Willinger, W., Govindan, R., Jamin, S., Paxon, V., and Shenker, S. (2002). Scaling phenomena in the Internet: Critically examining criticality. *PNAS* **99**, 2573.

[183] Yule, G. U. (1925). A mathematical theory of evolution based on the conclusions of Dr. J. C. Willis. *Phil. Trans. Royal Soc. London B* **213**, 21.

[184] Zachary, W. W. (1977). An information flow model for conflict and fission in small groups. *J. Anthropological Research* **33**, 452.

[185] Zhou, S. and Mondragón R. J. (2004). The rich-club phenomenon in the Internet topology. *IEEE Commun. Lett.* **8**, 180.

[186] Zlatić, V., Ghoshal, G., and Caldarelli, G. (2009). Hypergraph topological quantities for tagged social networks. *Phys. Rev. E* **80**, 036118.

Index